Introduction

The *Soil Conservation Technical Handbook* is a comprehensive collection of know how about soil conservation in world. Information has been gathered from individual knowledge and personal notes along with past often unpublished, or scarce copies of, printed material. Some of this material is derived from internal regional council manuals and papers.

The main forms of erosion in globe are covered in this Technical Handbook.

- **Mass movement erosion** – which occurs when heavy rain or earthquakes cause whole slopes to slump, slip or landslide. Most hill slopes steeper than 15 degrees are susceptible to mass movement, and those steeper than 28 degrees generally have severe potential. Storms are the primary triggers. This is the most common form of erosion in the hill country.
- **Fluvial erosion** – which occurs when running water gouges shallow channels or deeper gullies into the soil. On sloping land the gullies can cut deep into the subsoil or undermine surrounding soils.
- **Surface erosion** – which occurs when wind, rain or frost detach soil particles from the surface, allowing them to be washed or blown off the paddock. Surface erosion can occur on any land which is exposed to wind and rain but occurs largely outside the hill country.
- **Sediment erosion** – activities involved in earthworks, plantation forests, cropland and pasture management may all result in significant sediment loads being mobilised and often entering watercourses.

The Handbook is divided into two main parts. Part A covers the forms and processes of the main types of erosion in world. Readers can expect to gain a working understanding of erosion types, their typical geographic locations, as well as the overall principles of control. Descriptions are written from a practical point of view. More technical descriptions can be found in the references provided. A table at the end of each chapter summaries the main control technique headings and is cross-referenced to Part B.

Part B describes the control techniques. In most cases the Handbook will provide enough information for the practitioner to put together a soil conservation programme. Where more details (mostly engineering) are needed readers are referred to more specialised text. Often techniques apply to more than one erosion type and situation, for example maintaining good pasture cover is important in many situations. In these cases, general guidelines are given, as it is recognised that regional conditions will often dictate specific actions required.

The Handbook has been formatted so additional information can be added at anytime. It is anticipated that additional written material will be prepared in the future on associated land management topics and issues (for example indigenous biodiversity) as the demand requires. Space has also been provided for personal notes.

Contents

Part A
Principles

Part B
Practices

Part A
Principles

Chapter 1

Wind Erosion

1.1 INTRODUCTION

Wind erosion is the detachment and transportation of soil particles by wind when the airstream passing over a surface generates sufficient lift and drag to overcome the forces of gravity, friction and cohesion. Once a particle has been dislodged from the surface, it may be transported in suspension or by saltation or by surface creep. In high country, soil particles are often dislodged by needle ice formation and then removed by wind.

"Wind erosion, unlike soil erosion by running water, is often a rather subtle process, despite the quantity of sediment removed. Physical indicators of wind erosion, such as sand shadows behind obstructions and sediment ripples in fields, tend to be less obvious to the observer than the rill channels or large gullies cut by fluvial processes. Loss of topsoil by wind erosion over a relatively short time period can significantly decrease soil fertility and crop yield. Plant nutrients bound to soil colloids and organic matter can be removed from fields within small soil units (aggregates <0.2 mm diameter) that are readilytransported by wind" (Nickling and FitzSimons, 1985).

Wind erosion can also downgrade air quality and contribute to air pollution. Dust particles in the range of 0.25 to 10.0 microns aerodynamic diameter canoriginate from farms and be carried up to 10,000 metres elevation in the regional airmass. The Table 1.1 gives a useful overview of some of the effects of wind erosion.

1.1.1 Impacts on land use

For wind to erode soil, the following conditions need to be present:

- poorly aggregated soils
- surface soils that are loose, dry and finely divided
- wind that is strong enough to move soil.

In general, the finer and drier the soil, the easier it is eroded by wind when exposed through lack of vegetative cover. In New Zealand, wind erosion can occur from seashore to mountaintops. Areas exposed to either coastal winds (prevailing westerly or föhn winds the nor'westers), have been known to erode severely while soils were exposed. Basher and Painter (1997) cite background rates <0.2 t/ha/yr, extreme storm event losses >3000t/ha, and "average" storm losses ranging from 20–125 t/ha.

Table 1.1. Some physical and economic effects of wind erosion (after Wilson and Cooke, 1980)

Physical Effects	Economic Consequences
Soil Damage	
1 Fine material, including organic matter, may be removed by sorting, leaving a coarse lag.	1, 2, 3 Long-term losses of fertility give lower returns per hectare.
2 Soil structures may be degraded.	
3 Fertilisers and herbicides may be lost or redistributed.	3 Replacement costs of fertilisers and herbicides.
Crop Damage	
1 The crop may be covered by deposited material.	1-6 Yield losses give lower returns.
2 Sandblasting may cut down plants or damage the foliage.	1-4 Replacement costs, and yield losses due to lost growing season.
3 Seeds and seedlings may be blown away and deposited in hedges or other fields.	
4 Fertiliser redistributed into large concentrations can be harmful.	
5 Soil borne disease may be spread to other fields.	5 Increased herbicide costs.
6 Rabbits and other pests may inhabit dunes trapped in hedges and feed on the crops.	6 Increased pest control costs
Other Damage	
1 Soil is deposited in ditches, hedges, and on roads.	1 Costs of removal and redistribution.
2 Fine material is deposited in houses, on washing and cars, etc.	2, 3 Cleaning costs.
3 Farm machinery, windscreens etc. may be abraded, and machinery clogged.	
4 Farm work may be held up by the unpleasant conditions during a blow.	4 Loss of working hours and hence productivity declines.

These losses result in soil quality deterioration. In South Canterbury, Cuff (1981) estimated the following losses if 5 mm of soil depth is lost; this would be equivalent to 50 t/ha:

Active wind erosion, Hakataramea *(Photo: MWD)*

- nitrogen – 175 kg
- phosphorus – 45 kg
- potassium – 500 kg
- magnesium – 325 kg
- calcium – 525 kg

Since shallow soils retain less moisture than deeper soils, the cost of erosion may also be expressed in long-term production losses. Hayward (1969) gives examples of chou moellier crop yields in North Otago on soils of different depths:

- for topsoil depth of 300–375 mm, yields averaged 55 t/ha.
- for shallower soils near hilltops with depths of 75–225 mm, yields averaged 20 t/ha.

Benny and Stephens 1985 reported soil losses of 11-41 tonnes/ha/year on unprotected soils in South Canterbury, compared with 1.5–1.8 tones where conservation measures were practised. Wheat yields on the unprotected sites averaged 3.1 tonnes/hectare compared with 3.7 tonnes on the protected sites. Basher (1997) provides more detailed figures on erosion rates and areal extent for all erosion types occurring in the Canterbury Region, including wind.

1.2 FORMS, PROCESSES AND EXTENT OF WIND EROSION

1.2.1 Forms and processes

When wind blows across bare ground, grains of soil or rock are eroded if the stress exerted on them by air turbulence exceeds the forces that resist movement – that is, gravitational stress, together with friction and cohesion amongst the surface particles. Once entrained in the flowing air, grains are transported so long as the stress exerted on them by wind exceeds the force exerted by gravity. If wind stress declines – due to a drop in wind velocity – or if grains' momentum reduces – due to particle collision, or passage through vegetation – they are deposited.

The physics of the process constrains wind erosion to places where velocity and turbulence are high: exposed coasts; plains or broad valleys; and mountain ridges.

Wind erosion occurs only in places where soil is susceptible. Clay is rarely eroded by wind. Although its particles are extremely small, they have high cohesion. At the other extreme, gravel or stone fragments are rarely eroded. Although they lack cohesion, their particles are sufficiently large for the force of gravity to outweigh all but the most extreme wind stress.

Silt grains are easily eroded from topsoil by wind, because their particle size is very small and their cohesion is moderate. Sand grains are also eroded from topsoil, from beaches and from riverbeds, because their particle size is still sufficiently small to be entrained by wind, and their cohesion is low.

There are three characteristic forms of wind erosion:

- Wind blow: Just a millimetre or two of silty or sandy topsoil is stripped away, leaving no perceptible erosion scar. The transport process is visible as a cloud of dust. Silt or sand eventually settles downwind, many kilometres away. If repeated and persistent, deposits accumulate around the stems of growing vegetation as a thick even blanket of new soil.
- Wind scabs (deflation surfaces): Several centimetres of topsoil are stripped away by repeated wind blow, from a patch of ground several square metres in extent. The surface of the patch may be more resistant subsoil or a "lag" of gravel and stones left behind after fine particles have been blown from the topsoil. When large volumes of soil are eroded from scabs, much may be deposited a short distance downwind, forming drifts up to a metre high piled against obstructions like fences, shrubby vegetation, or hill slopes.
- Dunes: Many metres depth of sand is blown across the landscape leeward of beaches, as a result of alternate erosion of grains from the dune's windward side, of transport across its crest, and of deposition on the dune's leeward side. Different forms of coastal sand dune are described in Chapter 7, Sand Dune Stabilisation. At some places in New Zealand's inland landscape, thick deposits of river silt or volcanic ash have also been shaped into "fields" of small dunes, individually a few metres high and long.

For greater detail, see Bagnold (1941) for a definitive account of the physics of wind erosion, Painter (1978) for an account of the New Zealand landscape's susceptibility to wind erosion, and Cooke and Warren (1973) for comprehensive descriptions of the geomorphology.

1.2.2 Wind erosion research

Wind erosion has been the subject of research in the USA since the 1930s (see Fryrear et al's 2000 study). The first Wind Erosion Equation (WEQ) to quantify wind erosion was published by Woodruff and Siddoway in 1965. In 1996, Shao et al developed a physically based Wind Erosion Assessment Model (WEAM) to estimate erosion event movement across continents. A more process-oriented Wind Erosion Prediction System (WEPS) is under development (Hagen, 1991). Fryrear and colleagues describe a Revised Wind Erosion Equation (RWEQ) based on the parameters wind force, surface roughness, soil wetness, and crop residue or crop canopy.

Fifty years of wind erosion research by the USDA, Agricultural Research Service at Kansas State University was celebrated with a symposium in 1997, the proceedings of which may be accessed at www.weru.ksu.edu/symposium/proceed.htm.

In New Zealand, the problem was recognised as early as 1858 by Ferdinand von Hochstetter in his book, New Zealand. Its physical geography, Geology and Natural History. The first attempt to quantify soil losses resulting from wind erosion was that by Campbell (1945), who used a portable blower on a number of North Island east coast soils. He varied the speed of the blower to simulate soils under pasture and cultivation. The values of soil loss ranged from nil to 1668 t/ha/yr.

The forms and processes of wind were first described by Cockayne (1913) for sand country, and by Cumberland (1944) for arable land and depleted tussock rangeland. Much of the research into forms and processes has been undertaken in sand country; well-known work includes that of Brothers (1954), Cowie (1967), Hicks (1975) and Hesp (1990). Some research into the geomorphology of wind erosion on arable land is contained in unpublished reports and theses by Painter (1977, 1978) and his students. More detailed investigations have been carried out in recent years by Basher et al (1997, 2000).

A major review of erosion risk on arable soils (Ross et al, 2000) has been prepared for Environment Canterbury and is an essential reference on this topic.

In New Zealand's environment, the major factors affecting the occurrence of wind erosion, and its extent and severity, are:

- climate (wind patterns, precipitation, frost action)
- soil (texture, moisture, structure, organic matter content)
- topography (exposure, elevation, terrain roughness, localised funnelling of wind)
- cultural practices (cultivation, vegetation depletion).

Looking at these factors, it is clear that the near-surface soil water content is an important influence on wind erosion rates in New Zealand, where the evaporative energy supply can be very high and can rapidly reduce the near-surface water content. The soil's resistance to evaporative flux, combined with soil erodibility (for the predominantly light textured soils of New Zealand) determines whether a soil will erode.

1.2.3 Extent in New Zealand

New Zealand Land Resource Inventory (NZLRI) survey assessed wind erosion (NWASCO, 1974-79). Data from this survey were summarised by Eyles (1983) and are reproduced in the following table. The distribution of wind erosion as mapped in the same survey is shown in Appendix II.

Table 1.2. Erosion rankings, North and South Islands (after Eyles, 1983)

	Erosion Ranking	**Area (ha)**	**Area (%)**
		North Island*	
1	slight	255,500	48.5
2	moderate	131,700	25.0
3	severe	67,400	12.8
4	very severe	16,500	3.1
5	extreme	55,800	10.6
		North Island**	
1	slight	1,620,900	77.9
2	moderate	388,600	18.7
3	severe	58,000	2.8
4	very severe	12,000	0.6
5	extreme	1,500	trace

* Areas are those of map units in which wind erosion was recorded, with the given ranking.
** Areas are those of map units in which wind erosion was the dominant type.

In 1984, Salter used NZLRI data to undertake a more detailed review of the distribution and severity of wind erosion in New Zealand. He found that approximately 13 percent (3.4 million hectares) was affected. Wind erosion affects 4.6 percent of the North Island, mainly in three environments:

- the mobile, coastal dunes on the west coast of Northland and the Rangitikei-Manawatu that have not yet been afforested or have poor grass cover
- high-altitude (>700 m), volcanic ash-mantled, slopes in the central North Island that have poor vegetation cover due to frequent strong winds and cool temperatures

- low-altitude, loess-mantled argillite hill country and alluvial terraces in the eastern North Island with low-fertility soils and severe seasonal soil moisture deficits.

In the South Island, wind erosion affects 19 percent of the land, mostly on loess-mantled terraces and slopes in low-rainfall, seasonally dry eastern regions subject to frequent strong föhn winds:

- extensive alluvial plains in Canterbury, Marlborough and Southland with silty alluvial soils on younger terraces and loess soils on older terraces, that are susceptible to wind erosion when cultivated
- loess-mantled downlands in Canterbury and North Otago, subject to wind erosion when cultivated
- large inland basins in central Otago, Canterbury and Southland with loess-mantled terraces and moraines where extensive grazing by sheep and rabbits has led to severe vegetation depletion
- the steep, dry mountain lands of Canterbury, Otago and Marlborough with severe summer soil moisture deficits and widespread vegetation depletion
- hill country with shallow soils, discontinuous loess cover, severe summer soil moisture deficits and localised vegetation depletion in North Canterbury
- the exposed rolling uplands of Otago where vegetation is depleted.

According to the NZLRI survey, the area potentially subject to wind erosion is much larger than the area already affected. Approximately 27 percent of New Zealand being susceptible: 7.1 million hectares of the North Island, and 5.8 million hectares of the South Island.

1.3 PRINCIPLES OF CONTROL

The techniques available to control wind erosion have one or more objectives in common:

- reduce the cause of erosion (eg, reduce local windspeeds)
- enhance resistance to erosion (eg, reduce soil erodibility)
- reduce the undesirable consequences of wind erosion (eg trapping saltating soil, reducing physiological damage to plants).

In 1997, Basher and Painter review the extent to which these objectives have been attained. They note that wind erosion control in New Zealand has been dominated by the use of vegetative cover to reduce entrainment of soil particles or to provide a barrier which reduces wind velocity. There has been little use of tillage to increase surface roughness and produce non-erodible aggregates, in spite of advocacy by soil conservators.

Baker (cited by Ross et al. 2000) similarly emphasised the various options now available to increase surface roughness and retain soil. Basher and Painter (1997) also consider that there is greater opportunity to use irrigation to increase surface soil moisture content, and reduce the hazard at times when soil would otherwise be susceptible to wind erosion. Mitigation of wind-borne dust emission, usually by spraying water or waste oil, from rural roads, construction sites, quarries and other industrial locations is increasingly practised.

1.4 APPROPRIATE CONTROL PRACTICES

Table 3 summarises wind erosion control practices currently used on arable land, and at susceptible sites in the hill country and mountain lands. Control techniques include:

- shelterbelts
- pasture species and management
- cultivation and drilling practices
- stubble management
- adhesives and tackifiers
- irrigation.

For wind erosion control practices on sand country, refer to Chapter 5

Table 1.3. Summary of management practices for control of wind erosion.

Control Practice	Factors to Consider/ Technique		Refer to Part B for Practice Description
Shelterbelt	Siting		14.1
	Design		14.2
	Establishment		14.3
	Design description (various)		14.4
Pasture management & species	Revegetation	- Lowlands	4.1
		- Hill country	4.2
		- Mountain lands	4.3
	Pasture management	- Lowlands	2.1
		- Hill country	2.2
		- Tussock	2.3
Cultivation & drilling practice	Minimum tillage		1.3

	Direct drilling	1.4
Stubble management	Stubble retention	1.5
Others - adhesives & tackifier - irrigation		

1.5 REFERENCES

Anon. 1971 Wind erosion control. Soil & Water 8:23-27 and 29

Bagnold, R.A. 1979: Sediment transport by wind and water. Nordic Hydrology, v10, pp309-322.

Baker, CJ and Ross, CW 2000: The practical issues of increasing the adoption of no tillage in Canterbury, pp 114-119 in "Management of Erosion Risk on Arable Soils. A Review". Report No. U00/49, prepared for Environment Canterbury.

Basher, LR and Painter, DJ 1997 Wind Erosion in New Zealand. Paper presented at the conference "50 Years of Wind Erosion Research by USDA – ARS at Kansas State University". See www.weru.ksu.edu/symposium/proceed.htm

Basher, LR 1997: A review of erosion studies in Canterbury. Report produced for Canterbury Regional Council.

Basher LR, Hicks DM, Handyside B and Ross CW, 1997: Erosion and sediment transport from the market gardening lands at Pukekohe, Auckland, New Zealand. Journal of Hydrology (NZ) 36(1) p 73-75.

Benny, L.A., and Stephens, P.R., 1985: The feasibility of determining the influence of arable land management on topsoil depth. Soil Conservation Centre, Aokautere, Publication No. 7.

Brothers, N. 1954: Geology of South Kaipara Head. Unpublished Ph. D. thesis, University of Auckland.

Campbell, D. A., 1945: Soil Conservation Studies applied to farming in Hawkes Bay Part 1 Investigations into runoff and soil loss, NZ J. of Science and Technology 26(6): 301-332.

Cockayne, L. 1909: Report on the sand dunes of New Zealand. Appendix Journal House of Representatives, C-11.

Cooke, R. U. and Warren, A. 1973: Geomorphology in Deserts. London. Batsford.

Cowie, J. D. et al 1967: Soils of the Manawatu-Rangitikei sand country. Bulletin 29, Soil Bureau, DSIR.

Cumberland 1944 Soil Erosion in New Zealand – A Geographic Reconnaissance. Publ, Whitcombe & Tombs Ltd., New Zealand. Second Edition 1947.

Eyles, G.O., 1983: The distribution and severity of present soil erosion in NZ, NZ Geographer 39(1):1-28

Fryrear D W, Bilbro, JD, Saleh, A, Schomberg, H, Stout, JE and Zobeck, TM 2000: RWEQ: Improved Wind Erosion Technology. J Soil and Water Conservation, 55 (2) 183-189.

Hayward, J. A., 1969: The use of fractional acre plots to predict soil loss from a mountain Catchment. Lincoln Papers in Water Resources 7, New Zealand Agricultural Engineering Institute, Lincoln College 93 p

Nickling, W.G. and FitzSimons, J.G. (1985): Relationship of soil type and agricultural systems to wind erosion in Southwestern Ontario.

NWASCA Miscellaneous Publication No. 59 (Ed:. Sturrock 1984).

Painter, D.J. 1977: Wind erosiveness summaries for New Zealand sites. Report PR15, NZ Agricultural Engineering Institute, Lincoln.

Painter, D.J., 1978: The Pastoral Scene. Soil Erosion Rates on New Zealand Farm Land. Proceedings of the Conference on Erosion Assessment and Control in New Zealand.

Ross, CW, L Basher, CJ Baker 2000: Management of erosion risk on arable soils: a review. Landcare Research Report LC9900/090 prepared for Environment Canterbury regional Council. 119 p.

Salter, RT 1984: Wind Erosion. Pp 206248 in Speden, I. And Crozier, MJ (compilers) Natural Hazards in New Zealand. New Zealand National Commission for UNESCO, Wellington.

Soil Conservation and Rivers Control Council Circular SC 1971/16

Wilson, SJ and Cooke, RU (1980): Wind Erosion, pp 217-251 in Kirby, MJ and Morgan, RPC (Eds) Soil Erosion, Publ, J Wiley & Sons Ltd.

Chapter 2

Sheet and Rill Erosion

2.1 INTRODUCTION

Running water removes soil and organic matter from gentle sloping land and steeper sites by a variety of processes which often start off as splash erosion leading to sheet wash and as the conditions change this will lead into rill erosion which in time on the deeper soils – that is, loess or volcanic material – maydevelop gullies. This chapter will concentrate on sheet and rill erosion of soil by overland flow of water, on arable flat-rolling land, downlands and lower foothills. Sheet erosion may often coexist with rill erosion, and one may lead into the other.

2.1.1 History of sheet/rill erosion research

Despite the concern expressed about the magnitude of erosion and its impact on productivity, there have been few quantitative studies of surface erosion on arable land in New Zealand. Campbell (1945) showed that surface runoff as a percentage of rain under good grazed pasture was 42.2 percent compared to 0.3 percent for forest. Benny and Stephens (1985) estimated South Canterbury soil losses on a 12-degree slope of 40 t /ha/yr under annual cultivation, 23 t/ha/yr with ploughing one year in three, 3 t / ha/yr with continuous cropping and stubble mulching, and 0.4 t / ha / yr under permanent pasture.

Hunter and Lynn (1988, 1990) made measurements after storms in Canterbury. Basher (1997) reviewed the soil losses in Canterbury and identified high rates during storms of up to 11 mm through sheet and rill erosion. For cropping paddocks, there was considerable redistribution of the soil within paddocks as the soil was initially moved off slopes by rill erosion and deposited in swales, leaving a great variation of topsoil depths. There was a greater percentage of shallow topsoils under cropping than pasture, sufficient to impact on crop yields.

Basher et al (1997) published a summary of soil loss measurements at Pukekohe. This study showed that of the soil eroded within a paddock, a significant proportion was redeposited on lower slopes while a smaller proportion was delivered to streams. (Thompson and Basher 1999) identified some interesting factors due to an intense storm event at Pukekohe on 21 January 1999 where over 135 mm rainfall fell in four hours (exceeding the 1 in 100 year event). The soil was prone to downslope movement where slopes exceeded 4°. Paddocks steeper than 2° average slope which had seedbed preparation (eg, by rotary hoeing) had erosion rills, predominantly along wheel tracks, leading to more compound rill and sheet erosion further down the paddock as the runoff water increased in velocity. Erosion rates in some fields reached values as high as 600 t/ha. This meant an average loss of about 5 cm over the field, mainly due to rill erosion (sheet was considered only minor form); however, most soil was redeposited in nearby hollows or roadside drains.

2.1.2 Impacts on land use

Surface erosion can cause reductions of one to three-fifths in crop yield, two to four-fifths in pasture growth and up to nine-tenths in tussock biomass. Most of these production losses are reversible with appropriate soil conservation management.

In dryland areas like North Otago, Canterbury and Marlborough, soil losses on cultivated paddocks is still a regular occurrence. In North Otago between January and March 1986, it was estimated that 240 ha on 24 properties lost an average of 50 t/ha of fertile soil (5 mm depth of topsoil). This equates to 12,000 t of topsoil, which would require fertiliser estimated to cost $300,000 (1986 figures) to replace the major nutrients lost (Otago Regional Council, 1989).

Cuff (1978) estimated that typical nutrients lost after 10 mm topsoil is eroded equated to:

Total Nitrogen	350kg/ha
Total Phosphorus	90 kg/ha
Total Potassium	1000 kg/ha
Total Magnesium	650 kg/ha
Total Calcium	1050 kg/ha

2.2 TYPES, PROCESSES AND EXTENT

When rainwater runs off across bare ground, particles of soil or rock are eroded if three conditions exist:

- The soil's infiltration capacity is exceeded, so that water can build up instead of soaking into the soil. This condition may be met on dry soil during intense rain, but it more commonly occurs where infiltration capacity has been reduced to zero by water soaking in and saturating the soil beforehand. A related and equally common situation, is where water seeps from saturated soil and flows across its surface – that is, infiltration capacity turns negative.
- The water layer is sufficiently thick for gravitational stress on it to exceed capillary forces holding water to the soil's surface, so that water starts to flow instead of merely ponding.
- The stress exerted on soil particles by laminar or turbulent flow of water, is greater than the forces resisting movement – gravity, together with friction and cohesion amongst the soil particles.

Once entrained by water, particles are transported so long as the force of flowing water is sufficient to keep them clear of the surface (suspension) or bounce them along it (saltation). If water stress declines, for instance due to a drop in velocity where slope gradient lessens, or due to a drop in mass where some water soaks into the ground, then particles are deposited. This also happens if particles' momentum is reduced, for instance where they encounter vegetation at the bare ground's edge.

Fan deposited at slope foot, below rilled wheel tracks, Pukekohe. **Photo: D Hicks.**

The physics of the process determines where surface erosion is found in nature:

- areas subject to intense rainfall
- soils with low infiltration capacity or which quickly saturate
- depressions or breaks in slope, where soil water can converge and seep out.

Surface runoff rarely has sufficient mass or velocity to transport boulder or stones. Transportation of gravel occurs only if runoff is voluminous and fast. Sands and silts are most commonly transported by surface runoff, as they are easily entrained by a low volume of runoff moving at a slow velocity. Clays are less commonly entrained, as their cohesion gives greater resistance; but once in motion, their small particle size ensures they are transported a long distance before settling.

Three forms are left behind:

- Sheet wash is the almost-imperceptible removal of a few millimetres of topsoil at a time from parts of a slope. The process is visible as discontinuous sheets of surface runoff. A thin layer of soil is usually deposited a short distance away, where runoff slows and soaks in.
- Scabs are distinct scars where several centimetres of topsoil are removed from patches of slope several square metres in extent, by thin broad sheets of coalescing surface runoff (sheet flow). A distinct layer of soil is deposited if

surface runoff encounters an obstacle, for instance dense grass or rising ground; but if not, soil is transported off the slope. It may be deposited at the slope foot if gradient lessens sheet flow's velocity sufficiently for particles to settle; if not, eroded soil will be delivered into the nearest permanent watercourse.

- Rills are long narrow miniature channels, where anything from 10 to 50 cm of topsoil is removed by surface runoff concentrated into thick narrow threads. Its volume and velocity being greater than sheet flow, rill flow usually transports eroded soil off slopes and into streams. Repeated erosion of the same rill may develop it into a gully (see Chapter 4, Gully Erosion).

Refer to Bagnold 1966 for a definitive account of the physics of surface erosion by water; Henderson 1983 for an account of the New Zealand landscape's susceptibility; and Leopold, Wolman and Miller 1964 (Chapters 8 and 10) for a comprehensive description of the geomorphology.

2.2.1 Extent

Surface erosion ie including sheet and wind (note it is difficult to separate soil loss from sheet and wind erosion) is found throughout NZ. NZLRI identified that 23.3 percent of the North Island and 74 percent of the South Island is susceptible to surface erosion. The following percentages indicate agricultural land that is susceptible to surface erosion processes:

- Otago 88 percent
- Canterbury 86 percent
- Nelson-Marlborough 77 percent
- Southland 59 percent
- Northland 41 percent
- Hawkes Bay 30 percent.

The trend drops to 6-9 percent in regions such as Taranaki, East Coast and Gisborne. where mass movement erosion predominates (Clough and Hicks 1993).

In the South Island, particularly east of the Southern Alps, loess is the most extensive soil-forming material. It underlies 7.45 million hectares–just over 51% of the soils (Raeside, 1964). However in some areas the amount of loess may be small; the stony soils on the Canterbury Plains are an example. In the North Island, the soils most susceptible are those formed from loess, volcanic ash, and pumice.

2.3 PRINCIPLES OF CONTROL

There are a number of interrelated factors, which can initiate and accelerate surface wash and removal of soil particles. Some common examples in New Zealand are:

- overgrazing by domestic farm animals (eg, due to large blocks with sunny and shady aspects)
- overgrazing by animal pests such as rabbits, hares, goats and feral animals

Localised rill erosion West Otago. Photo: Otago Regional Council.

- burning of native grassland/tussock and inter-tussock species
- burning of scrub and plant pests (eg matagouri, gorse and broom)
- burning of cereal crop stubble and removal of beneficial organic matter
- cultivation up- and down-slope and cultivating slopes over 20°
- cultivation at regular intervals growing similar crops/pasture
- compaction by stock and machinery and the formation of compaction pans in the soil
- which reduces normal water infiltration and increases overland flow
- cultivation of soil leaving a very fine tilth (i.e, with a high percentage of particles smaller than 5-10 mm).
- forest harvesting with skidder hauling machinery
- dozing steep slopes

Maintaining and improving ground cover to ensure that all the surface layer has living or decomposing vegetative material on it will help protect soil particles from rainfall events. Control measures become more critical on steeper land ie greater than 12° slope where soils are loessial, of volcanic origin or sandy in nature.

Principles of control fall into three groups:

- Firstly, there is a need to carry out water management to ensure that surface runoff water is removed safely and does not speed up or concentrate so earthworks such as graded banks, contour furrows, strip cropping and conservation tillage, soil drainage and floodbanks are constructed. aeration and Land Use Capability management.
- Secondly, it is important to ensure that a healthy, vegetative ground cover exists over the soil for as long as possible. Practices that maintain or improve the soil fertility, soil organic matter, soil flora and fauna as well as utilising the best plant species for the site (eg, drought grasses for dryland districts and water-adapted plants for wetter sites) should be adopted. This would include practices such as cultivation management, conservation tillage, soil drainage and aeration and Land Use Capability management.
- Thirdly, it is important to ensure that ongoing soil, plant and farm management practices are maintained or improved on all susceptible sites.

2.4 APPROPRIATE SOIL CONSERVATION PRACTICES

See Table 2.1 below.

Table 2.1. Summary of management practices for control of sheet and rill erosion

Type of erosion	Primary Soil Conservation Principle	Factors to Consider	Examples of Practices*
Splash and sheet erosion (collectively called sheet erosion)	Ensure effective ground cover, soil structure and health be maintained and improved. Create a stable seedbed.	Rainfall, storm frequency, infiltration capacity Topography and aspect Slope angle, slope length Geology, regolith (loess, ash) Soil type (depth, texture, structure, percentage of organic matter) Past tillage and vegetation type and rotations Land use – pastoral, cropping, horticultural, forestry or urban use Stock type	Water control Conservation tillage Drainage and aeration Earthworks Stubble mulching Rotational cropping Pasture sward maintenance Matching crop and pasture species to the site Timing cultivation when soil conditions are right (low risk of erosion) Adjusting grazing pressure to soil conditions i.e. avoid heavy stocking when risk of erosion is high

Rill erosion	Ensure effective ground cover, soil structure and soil health be maintained and improved. Create a stable seedbed.	As above (often on steeper slopes)	As above

*May use any one or a combination of them

Descriptions of the various practices for control of surface erosion are found in sections of Part B:

1. Cropland Management for Surface Erosion Control
2. Pasture Management for Surface Erosion Control
3. Fencing Management for Surface Erosion Control
4. Pasture Revegetation Practices
9. Runoff Control Practices - Lowland

2.5 REFERENCES

Bagnold, R. A. 1979: Sediment transport by wind and water. Nordic Hydrology, v10, pp 309-322.

Basher, L. R. et al 1997: Erosion and sediment transport from the market gardening lands at Pukekohe, Auckland. NZ Jour. Hyd., v36, pp73-95.

Basher, L.R., et al 1999: Assessment of erosion damage at Pukekohe during the storm of 21st January 1999, NZARM Broadsheet July 1999 pages30-31

Basher L.R., et al 1995: Surface Erosion Assessment in the South Canterbury Downlands, New Zealand using 137 Cs Distribution, Australian Journal Research, 33, 787-803

Benny, L.A., and Stephens, P.R., 1985: The feasibility of determining the influence of arable land management on topsoil depth. Soil Conservation Centre, Aokautere, Publication No. 7.

Campbell, D. A., 1945: Soil Conservation Studies applied to farming in Hawkes Bay Part 1 Investigations into runoff and soil loss, NZ J. of Science and Technology 26(6): 301-332.

Clough, P. and Hicks, D.L.,1993: Soil Conservation And The Resource Management Act Summary, Maf Policy Technical Paper 93/2 for Maf Policy Wellington

Eyles, G.O., 1983: The distribution and severity of present soil erosion in NZ, NZ Geographer 39(1):1-28

Haynes, R.J., 1995: Soil Structural Breakdown and Compaction in New Zealand Soils MAF Policy Technical Paper 95/5

Henderson, R. 1983: Values of the rainfall and runoff factor® of the universal soil loss equation for New Zealand. Internal Report 844, Hydrology Centre, Water and Soil Division, MWD.

Hicks, D.L., 1995: Control of Soil Erosion on Farmland, MAF Policy Technical Paper 54/4

Hunter, G. G. and Lynn, I. H. 1990: Storm induced soil losses from South Canterbury and North Otago downlands. DSIR Land Resources Technical Record 22, 7p.

Leopold, L. B., Wolman, G. M., and Miller, J. P. 1964: Fluvial processes in geomorphology. San Francisco. W. H. Freeman.

North Otago Sustainable Land Management Group 1995: North Otago Sustainable Land Management Guidelines, North Otago Sustainable Land Management Group in association with the Otago Regional Council

North Otago Volcanic Soils Environmental Group 1996: Sustainable Land Management Guidelines for the Waiareka Soils, North Otago Volcanic Soils Environmental Group in association with the Otago Regional Council.

NWSCA 1985: Soil Erosion in New Zealand, National Water and Soil Conservation Authority Publication no 49

NWSCA 1969: Land Use Capability Survey Handbook, Water and Soil Division Ministry of Works, Wellington, New Zealand

Painter, D.J., 1978: The Pastoral Scene. Soil Erosion Rates on New Zealand Farm Land. Proceedings of the Conference on Erosion Assessment and Control in New Zealand.

Raeside, J.D., 1964: Loess Deposits of the South Island, New Zealand, and Soils Formed on them. N.Z. Journal of Geology and Geophysics. 7: 8V-38

Shepherd, T.G., et al 2000: Soil Management guidelines for Sustainable Cropping, Landcare Research New Zealand

Thompson, T., 1999:Pukekohe Storm Event 21st January 1999, NZARM Broadsheet March 1999 pages 51-54

Watt, J.P.C., 1972: Loess soils and problems of land use on the downlands of the South Island New Zealand, Otago Catchment Board publication No 4.

Chapter 3

Shallow Mass Movement – Slip Erosion

3.1 INTRODUCTION

Slip erosion is a shallow and rapid sliding or flowing movement of the soil and subsoil, exposing a slip surface which is approximately parallel to the slope. Debris comes to rest in the area from the base of the exposed slip, to the toe of the slope. There can be some rotational movement, leaving a concave "slip surface" or failure plane. The failure plane is normally less than 1m (but sometime up to 2m) below the original surface.

It is most evident during or immediately after heavy rain. The basic reason for this is saturation of the soil, which increases the weight of the soil mass, lubricates the failure plane, and turns pore water pressure positive. The resisting forces are shear strength of the soil, cohesion of the material, and tensile strength of plant roots in the soil. The large scale clearing of forest cover from the 1870s on the hill country in New Zealand eventually resulted in widespread and accelerated slip erosion as well as other forms of mass movement.

This type of erosion is widespread on hill country throughout New Zealand. It is a natural process that occurs on all types of terrain and under all types of vegetation. Accelerated slip erosion may occur on unstable terrain when native forests are removed and replaced with grass, or between rotations of plantation forestry. Most slip erosion occurs during high intensity rainfall or prolonged wet periods when the regolith becomes saturated, pore pressures rise, and slope failure occurs.

Slip erosion is classified as either soil slip or earth slip in the NZLRI. (Eyles 1985) Treatment and recovery tend to be different for each type.

3.1.1 History of slip erosion control

Severe storm events like the Esk Valley floods in 1938 left extensive areas of hill country covered in shallow slips. Farmers realised that the poor quality of the pastures may have contributed to the slipping as well as the loss of forest cover. Rabbits were widespread; pasture management was almost non existent with few fences and almost no fertiliser having been applied. In the late 1940s experiments were made in several places throughout New Zealand with aerial seeding and the application of fertilisers. This eventually led to huge improvements in hill country pastures and a reduction in erosion (McCaskill, 1973).

However, it was shown that, in unstable areas, better pastures alone would not solve soil erosion. Poplars and willows were planted as unrooted poles into farmland in an effort to reduce soil loss and hundreds of thousands of poles were planted on hillsides in the following 50 years. Poplars were shown to have an extensive rooting system, and they could dry out the soil by transpiring moisture during spring and summertime. In recent years block planting with seedling trees has become more extensive. Pines, gums and wattles have been shown to have similar qualities in binding the soil together as well as being able to remove moisture from the soil throughout most of the year.

Heavy stock grazing on hill land susceptible to slip erosion can lead to pugging in wet weather, with rainwater stored in hundreds of depressions made by hooves. Continued saturation of the surface may lead to failure of the slope. Improved livestock management by subdivision and other grazing practices can help reduce the risk slip erosion.

The effective control of slip erosion requires good planning and recognition of areas where slips may occur. Often farmers will actively plant poles on sites that have already slipped. It is usually best to plant on sites that have a potential to slip and careful examination of an area can identify these places. Land use capability surveys are a key to identifying areas where preventative planting should be established.

3.2 TYPES, PROCESSES

Mass movements of soil or rock occur when stresses (downslope component of gravity pulling soil down the slope, pore water pressure, loading by vegetation, seismic waves propagating through the soil) exceed resistances (in-slope component of gravity holding soil to the slope, friction and cohesion of soil particles, reinforcement by vegetation roots). Slope failure occurs by three processes, acting singly or in combination: falls, slides, and flows.

The operation of three processes in varying combinations; differences in the soil properties that aid or resist failure; and the variety of triggering mechanisms – physical or chemical weathering,changes in pore water pressure during rainstorms or wet weather, undercutting or overloading of slopes, passage of earthquake waves – produce many different forms of mass movement. There is no standard nomenclature (despite several attempts at international standardisation).

3.2.1 Falls

Falls are a sudden rupture, followed by free fall of soil rock fragments, away from a near-vertical slope. The rupture is manifest as a fresh shear surface (failure plane) on the bluff or scarp, with a pile of loose debris (talus cone) beneath.

Earthslip erosion, Te Pohue. *Photo: Hawke's Bay Regional Council.*

Soil slip, Taranaki. *Photo: M. Tuohy.*

3.2.2 Slides

Slides result from elastic deformation, followed by sudden rupture and sliding of a soil or rock mass, down a steep slope. If movement is slight, the rupture is visible as tension cracks around the head and sides of a soil block that is otherwise intact. If movement is greater, the rupture is exposed as a sloping shear surface in the upper slope, with a disintegrated soil block (debris tail) farther downslope.

3.2.3 Flows

Plastic deformation, followed by interparticle shear and flowage of soil or rock mass, down a low-angle slope, forms a flow. If flow is slow, it manifests itself as surface rumpling of an internally deformed soil mass. If it becomes rapid, it is manifest as a disintegrated jumble of soil fragments (flow debris) deposited within downslope of the failure zone. A failure plane separates the soil mass or flow debris from intact regolith, but is rarely exposed. A flow usually induces some slide-type failures in the more rigid soil around its margins; these become more-or-less intact soil blocks rafted along in the flow debris.

3.2.4 Shallow mass movement in New Zealand

In New Zealand, a practical distinction is drawn on the basis of failure depth – whether the mass movement is shallow (in the soil) or deep (in underlying regolith). This separates mass movements which may be easily stabilised and routinely undertaken, from those where stabilisation is difficult but may need to be attempted (Chapter 4, Deep-seated Mass Movement- Slides, Slumps and Flows).

Four forms of shallow mass movement are common in the New Zealand landscape. The distinctions are again practical, differentiating two forms which can be controlled by vegetative techniques alone from two for which engineering measures are usually needed:

- Slips – These are shallow landslides in soil or weathered regolith on steep slopes (Figure 1 – photo). They are also found on low-angle slopes where regolith is susceptible to rupture and deeply-weathered rock with a high clay content. Occasionally found where steep slopes (otherwise stable) are mantled by volcanic ash or loess.
- Earth flows – These are shallow flows on low-angle slopes (Figure 2 – photo), where regolith is susceptible to plastic deformation, notably weathered mudstones which contain swelling clay minerals such as bentonite or montmorillonite. Earth flows are also found in colluvium (slip debris) on footslopes and occasionally found where low-angle slopes (otherwise stable) are mantled by weathered volcanic ash, loess or till.
- Debris avalanches are shallow, rapid landslides in regolith on upper mountain slopes (see Chapter 6 (Part A), Mountain Lands Erosion for description).
- Debris flows – shallow, rapid flows in colluvium (avalanche debris) on lower mountain slopes (see Chapter 6 for description)

Crozier 1986 gives detailed descriptions of the four forms, illustrated by many New Zealand examples. The physics of slope failure is complex, but essential to understanding the circumstances under which any control technique can (or cannot) work. Readers are referred to Carson and Kirkby (1972) (Chapters 5 to 7) for a definitive discussion of the topic. Scheidegger (1984) also provides a useful summary of the physics.

3.2.5 Extent of shallow mass movement

The NZLRI identified 7,351,600 hectares of land as susceptible to soil slip or earth slip erosion (refer to Appendix II). 1,044,800 hectares were identified as susceptible to earth flows (refer to Appendix II); most of this is terrain affected by shallow flows, though some deep-seated flows may be included. 2,821,900 hectares of land were identified as susceptible to debris avalanches from debris flows, Appendix II be taken as indication extent of these also.

3.3 PRINCIPLES OF CONTROL

Revegetation trials in Gisborne (Quilter, 1993) and in the Wairarapa (Lambert, 1993) have shown that it takes up to 20 years to restore productive pastures on erosion scars and this may only be 70-80 percent of dry matter production from uneroded ground. The successful introduction of pasture plants onto eroded sites requires retirement from grazing, annual fertiliser applications and careful stock management. The cost of remedial work may be uneconomic compared with other opportunities for farm investment. Protecting land from erosion is therefore a more attractive investment than repairing erosion after the event.

Earthflow, Hawke's Bay. *Photo: Water & Soil Division, MWD.*

3.3.1 Preventive works

The prevention of soil slip erosion usually requires a modification to pastoral farming practices. Susceptible areas on farms are readily identified by a succession of slip scars dating back to the time of development, and in many places even before forest clearance.

There are several options for reducing the risk of soil slip erosion. Improving pasture vigour by introducing deeper rooting species and regular applications of fertiliser will no suffice. It must be accompanied by:

- Spaced planting of trees in pasture, to provide the soil with root reinforcement, or
- Retirement from grazing, followed by planting with commercial timber trees, or
- Reversion to native scrub cover

3.4 APPROPRIATE CONTROL PRACTICES

Descriptions of the various practices for control of shallow mass movement are found in chapters/sections of Part B:

3.2	Fencing Management Practices for Erosion Control - Hill Country
4.2.	Pasture Re-Establishment - Hill Country
8.2. (8.3)	Managed Reversion of Retired Land - Hill Country
12.	Pole Planting

3.5 REFERENCES

Carson, M.A. and Kirkby, M.J. 1972: Hillslope form and process. Cambridge University Press.

Crozier, M. J. 1986: Landslides. 212pp, Croon Helm, London.

Eyles GO 1985: The New Zealand Land Resource Inventory Erosion Classification. Miscellaneous Water and Soil Publication Number 85, National Water and Soil Conservation Authority.

Lambert, MG, Costall, DA, Foote, AG, Trustrum, NA 1993 Proceedings of the New Zealand Grasslands Association 55, 177-181

McCaskill, LW 1973 Hold This Land, AH & AW Reed Ltd.

Quilter, S.J., Korte, C.J., Smithe DR. 1993) Proceeding of the New Zealand Grasslands Association 55, 187-191

Scheidegger, A.E. 1984: A review of recent work on mass movements on slopes and rockfalls. Earth Science Reviews, v21, pp225-249.

Chapter 4

Deep-Seated Mass Movement – Slides, Slumps and Flows

4.1 INTRODUCTION

New Zealand has some spectacular examples of deep-seated mass movements. Lake Waikaremoana and Lake Tutira in Northern Hawkes Bay were both formed by massive collapses of earth and rock into deep gorges. In recent times, the Abbotsford block glide in Dunedin demonstrated what happens when unstable slopes are modified for urban development. Less spectacular, but similar movements of soil, rock and vegetation regularly damage roads, railways, even houses, in many parts of the country. Huge investments in afforestation or structural works have been undertaken to reduce damage from such events.

Geology, water and gravity are the main factors in deep seated mass movement. The softer rocks of the Tertiary period, experience slumping and earth flows in many forms. Up to 41 percent of farmland in the Gisborne District is susceptible to earthflow erosion. Other examples of deep seated movement include, collapses of coastal cliffs, large slumps which occasionally form a temporary blockages in rivers and streams, and regional landslides where geological formations slowly collapse through an area several square kilometres in extent.

4.2 TYPES, PROCESSES AND EXTENT

The processes by which deep-seated mass movement occur are the same as those already described for shallow mass movement in Chapter 3. The distinction is one of scale – in addition to failing at depth in the regolith, deep-seated mass movements extend over many hectares of land; in extreme cases, several square kilometres. Their impact on land use is similarly large-scale. New instances of failure are uncommon, but the New Zealand landscape is dotted with historic (and prehistoric) failures where ongoing movement restricts use of the land for farming and forestry or threatens roads and railways.

Deep mass movements may be differentiated into three forms: slides (or block glides), slumps, and flows. The distinctions relate to feasibility of stabilisation, and what types of measure are worth attempting. Crozier (1986) gives further information about the three forms, illustrated by examples from the New Zealand landscape.

4.2.1 Slides (block glides)

A slide is a large block of soil and regolith, which moves sideways along a sloping shear plane in underlying rock. Most of the moving mass stays intact, though its edges break up into smaller blocks and fragments. The best-known example in recent years has been the Abbotsford slide, which left houses severely damaged at its upper edge where the sliding block separated, but more-or-less intact on other parts of its surface. Elsewhere in the country, it is not unknown for farming to continue on slowly moving block glides, so long as fences and tracks can be adjusted across their edges.

4.2.2 Slumps

Slumps are made up of several blocks of soil and regolith, which move in an arc above a curved shear surface (rotational failure plane) in underlying rock. The moving mass is separated from the slope behind by a high vertical rupture or "headwall". It breaks into an uneven surface of backward-tilted segments separated by lower ruptures or "scarps". Drainage is disrupted, and swamps or pond develop in depressions between each segment and scarp. A well-known example is the Ruru slump at Tinui in the Wairarapa, which is slow-moving and has been farmed for over 60 years.

Slump during Cyclone Bola. *Photo: Landcare Research.*

Deep earthflow, hill country cleared of bush, near Karitane, Otago. Photo: D Hicks.

4.2.3 Flows

Flows are made up of soil and regolith, which moves by inter-particle or inter-layer shear above a failure plane in underlying rock. The failure plane may be either planar or curved. The surface breaks into hundreds of hummocks, roughly aligned as arcuate ridges transverse to the direction of flow, and separated by tension cracks which form low scarps. Drainage is extremely disrupted. A typical example is the landscape either side of Highway 1 at Kilmog Hill, north of Dunedin.

4.2.4 Extent of deep-seated mass movements

The initial failure of a deep-seated slide, slump or flow is rapid and catastrophic. Many are triggered by earthquakes, although heavy rainfall or basal undercutting by streams are other common triggers. The mass movement debris semi-stabilises and revegetates, sufficiently for indigenous forest to establish within several hundred years. However, underlying regolith and drainage remain disrupted, creating the right conditions for secondary failure. Secondary failure usually happens at a slow rate, on patches of the debris rather than all of it, in some years but not every year.

Deep-seated slides, slumps and flows are widespread in New Zealand's tectonically disturbed landscape. The 109,900 hectares of terrain recorded as susceptible to slump erosion by NZLRI (Appendix II) is at best a partial indication of their extent; mapping recorded slumps on soft marine sediments but excluded deep-seated mass movements on the harder greywackes and metamorphic rocks. Institute of Geological and Nuclear Sciences (IGNS) has recently compiled a nationwide list of deep-seated mass movements greater that 1,000,000 cubic metres in volumes. The data used for the list have come from geological maps and inspection of aerial photographs (McSaveney et al 1998). The IGNS maps, though currently unpublished, are available on request.

4.3 CONSERVATION MEASURES TO CONTROL EROSION ON DEEP-SEATED MASS MOVEMENTS

Secondary failures, on the surface of existing deep-seated mass movements, may be stabilised by a mix of vegetative, soil drainage, and runoff control measures.

Stabilisation of primary failures is difficult and sometimes impossible. Engineering techniques are essential to success and have rarely been attempted in New Zealand's sparsely populated landscape – only at a few places, where deep-seated mass movements threaten lines of communication or urban areas. Consequently, there are no standard practices available for inclusion in Part B.

Practices are better-developed in several countries overseas, where the landscape is susceptible to deep-seated slope failures, and dense populations need to be protected. New Zealand's geotechnical engineers generally adapt the overseas practices, as and where needed here. Interested readers are advised to obtain information from Japan, Taiwan, Hong Kong, the United Kingdom, France, Canada or the United States; these being the countries where relevant engineering practice appears most advanced.

4.4 APPROPRIATE CONTROL PRACTICES

Table 4.1 Control techniques for deep-seated mass movements on different geomorphology.

Erosion Type	**Geomorphology**	
	Crushed Argillite	**Other Sedimentary Rocks**
Flows	Erosion control forestry Dewatering Debris dams	Pole planting Dewatering
Slides	Dewatering Pole planting	Dewatering Pole planting
Slumps	Erosion control forestry Dewatering	Erosion control forestry Dewatering Pole watering

Descriptions of the various practices for control of deep-seated mass movement are found in chapters/sections of Part B:

3.2	Fencing Management for Erosion Control Đ Hill Country
8.2. (8.3)	Managed Reversion of Retired Land Đ Hill Country
10	Dewatering Techniques for Deep-seated Mass Movements
12	Pole Planting
13	Erosion Control Forestry

4.5 REFERENCES

Crozier, M. J. 1986: Landslides. 212pp, Croon Helm, London.

Hicks, D. L. 1995: Control of Soil Erosion on Farmland, MAF Policy Technical Paper 95/4

Maunsell, J.R., Slump Control, FP45, Unpublished report held by Wellington Regional Council, Masterton

Painter, D J 1973 Prevention and Control of Mass Movement and Gully Control, NWASCO 1973

Smith, R K. 1973 Prevention and Control of Mass Movement and Gully Control, NWASCO 1973.

Chapter 5

Gully Erosion

5.1 INTRODUCTION

Gully erosion is "the removal of soil or soft rock material by water, forming distinct narrow channels, larger than rills, which usually carry water only during and immediately after rains" (Bates and Jackson 1980).

Tunnel gully erosion is "a compound erosion form initiated by the subsurface concentration and flow of water, resulting in scouring and the formation of narrow conduits, tunnels or pipes. Soluble, dispersive or low strength material is removed, ultimately resulting in collapses, visible either as 'holes' in the land surface, or as gullies developed by the collapse of pipes – followed by continued erosion" (Eyles 1985). Tunnel gullies are also called "tomos"'or "underrunners".

5.1.1 History of gully control

Many gullies occur naturally in the New Zealand landscape. They are often formed as a result of geological phenomena such as volcanic eruption or faulting and crushing of marine sediments. Over long periods of time, these natural gullies often become stabilised as the channel reaches a stable gradient, the side walls of the gully achieve a stable slope, and the natural vegetation becomes well established.

During the nineteenth and twentieth centuries, when large areas of land were cleared for pastoral development, the changes in runoff patterns and areas of exposed ground often resulted in old gullies reactivating, and new gullies being initiated. Sometimes the erosion occurred quickly, immediately following the development phase. In other situations, a number of years or even decades passed before the erosion became evident.

Following the enactment of the Soil Conservation and Rivers Control Act 1941, the Soil Conservation and Rivers Control Council (Soil Council) was established, and catchment boards were set up around New Zealand to promote soil conservation, prevent or mitigate soil erosion, and prevent damage by floods.

"In the early days of the Soil Council, soil conservation in the southern half of the North Island meant gully control" (McCaskill, 1973).

While the principles of gully control were reasonably well understood, the early soil conservators had to develop specialised practices for different parts of New Zealand (often by trial and error) to achieve success. From the outset, it was recognised that vegetative protection (trees and ground cover) provided the main long term method of control.

A historical account of gully control practices, initially developed in Soil Conservation Reserves before being used for subsidised soil conservation work on private land, is given by McCaskill (1973)

5.2 TYPES, PROCESSES AND EXTENT

A gully is a distinct channel, carved into a hillslope or valley bottom by intermittent or ephemeral runoff. Such channels are carved where the force exerted by flowing water – a function of its mass and velocity – exceeds the subsoil's resistance – a function of gravity (holding it to the slope) together with properties of its surface particles (binding them together, and resisting the tractive stress of water). (A definitive account of the physical processes which form gullies can be found in Leopold et al, (1964), Chapter 10-11.)

The processes – removal, transport and deposit of particles by flowing water – are exactly the same as for surface erosion and stream erosion. What makes gully erosion distinct is the scale and frequency at which processes operate:

- sheetwash or rilling – in topsoil; ephemeral during rain
- gullying – in subsoil; intermittent during and after rain
- stream bed scour and bank collapse – in alluvium or weathered rock, continuous but slight during low flows; intermittent but severe during floods.

There are four main forms of gully, although some gullies may be a combination, gully erosion often occurs either in association with one or more other erosion types. These types are U-shaped, deep-seated, tunnel and mountain gullies.

5.2.1 U-shaped gullies

U-shaped gullies are formed in loose to non-cohesive, uncompacted lithologies. They are initiated by channel gullying and exacerbated by waterfall gullying and even. They tend to form a box or U shape.

U-shaped gullies develop in clearly identified stages. When overland flow occurs during or following rainfall, stormwater runoff is concentrated into depressions. Rill erosion is initiated when there is downward scour of the topsoil. This may occur where the surface is less resistant to particle detachment or where flow velocity is increased by an abrupt gradient change. Rills deeper than 60 cm deep and wider than 30 cm are classified as gullies (Brice, 1966).

As the downward cutting continues, a gully head forms and recedes upstream; as the gully deepens and widens there is lateral erosion of the gully sides. A waterfall may develop at the gully head where overland stormwater flow plunges from the ground surface into the eroding channel. Sometimes there are a series of small waterfalls in the channel itself. The depth to which a gully will erode depends on the subsoil. Downward cutting will continue until a more resistant layer is reached.

U shaped gully on alluvial tephra deposits. *Photo: Environment BOP.*

5.2.2 Extent of U-shaped gullies

U-shaped gullies occur on weathered recent sands, on unconsolidated (alluvial) Pleistocene sands, on Quaternary tephra deposits, and on gravels. This form of erosion is widespread throughout New Zealand where soils and soil structure are susceptible to fluvial erosion. It will often occur when surface vegetation is in poor condition and stormwater runoff is concentrated.

U-shaped gullies commonly occur on loose, non-cohesive or uncompacted soils such as loess, pumice, volcanic ash, alluvial sands and gravels. U-shaped gullies may form on very flat slopes once the gully process is initiated by a small abrupt gradient change or nick point. Trials at Rerewhakaaitu have demonstrated that the gully erosion process can be initiated on unconsolidated exposed ash soils where slopes are greater than 7°.

5.2.3 Deep-seated gullies (V-shaped gullies)

Deep-seated gullies form where hillslope watercourses cut through surficial regolith into unstable rock beneath. They are common on three kinds of hill country:

Soft marine sediments: Here the process is entirely physical. The sediments are at best weakly cemented by calcite and are often just indurated (compacted). Where vegetation is removed from an ephemeral watercourse – for instance by land clearance forfarming or forestry – there is no obstacle to entrainment of sediment grains by running water. The watercourse cuts a deep trench headwards and its over-steepened sides collapse. Sediment is delivered into the permanent stream at the slope foot and re-worked along its course through the valley bottom. If the quantity of sediment is large, it aggrades as a fan (at the gully mouth) or a terrace (in the valley bottom). Such gullies usually stabilise as narrow, deep slits in a hillslope. Occasionally they develop into branching networks that dissect the entire slope.

Cemented marine sediments: This process is both physical and chemical. Alternate wetting and drying disaggregates weathered rock exposed in the bed of an ephemeral watercourse. Calcite cement is leached by weakly acid soil water, percolating towards the bed. Once calcite bonding is lost, any swelling clay minerals (bentonite and montmorillonite) expand. Leaching and swelling may be accelerated if pyrite (small quantities are present in some marine sediments) renders soil water more acid than normal. The watercourse begins cutting into the loosened rock; as it cuts down, fresh rock is exposed to physical and chemical weathering. Downcutting often triggers slips or earthflows around the gully's sides and head.

If such a gully is stabilised by vegetation while incipient, it remains narrow and V-shaped. If not, it develops into a branching network of channels incised tens of metres into "badland" terrain, several hectares in extent.

Crushed rocks: Where rocks have been crushed by movement along faults or by intense folding during uplift, soil water can percolate deep into the crushed rock and weathering is more rapid. If a watercourse crosses such a crush zone, it is eroded much more easily than the hard rock to either side.

Such a gully may remain small, deep and narrow where hard rock downslope bars downcutting or hard rock upslope constricts backcutting. Where a crush zone extends through a slope, the gully may extend over several hectares and coalesce with a similar feature on the other side of a ridge.

There is no single New Zealand publication which gives a comprehensive account of deep-seated gullies. Some relevant descriptions are found in reports by Gage and Black (1979), De Rose et al (1998), and Phillips, Marden and Miller (2000).

5.2.4 Extent of deep-seated gullies

These gullies occur throughout New Zealand on steep hill country or colluvial slopes. In the North Island, they are confined largely to crushed argillite, jointed mudstone, and jointed fine siltstone in association with earthflow and soil slip erosion forms. In the South Island, most deep-seated gully erosion is on greywacke and sub-schists on colluvial slopes, in association with sheet and scree erosion.

V shaped gullies on crushed argillite country. **Photo: Gisborne DC.**

V shaped gullies can occur in the following situations;

- as finger gullies in Tertiary and lower Cretaceous deposits
- on mudstone hill country in association with soil slip, slump, and earthflow erosion
- on crushed argillite hill country as bifurcated or amphitheatre gullies
- On greywacke or schist hill country as shallow talus colluvial material, deeper greywacke which may have been structurally altered or have a high argillite content, or as weathered greywacke.

5.2.5 Tunnel gullies

These gullies are formed in variable lithologies from a concentration of underground water flow above a less permeable soil layer. The gully process is initiated by underground 'piping' as lateral flow of soil water entrains particles. The pipes enlarge to form tunnels that may ultimately collapse to form gullies.

Tunnel gullies form in situations where there is a variation in permeability within the soil profile, and water is able to infiltrate into the subsurface layers.

Subsurface water is concentrated where there is non-cohesive or soft lithology overlying more resistant layers. The flow of water along the resistant layer initiates 'piping', where the soils are soluble, dispersive or prone to fluvial erosion. The 'piping' erosion process forms tunnels. As the tunnels enlarge, they can collapse to form holes or continuous gullies as surface erosion features.

Following collapse of the tunnel gully, the normal gully erosion process continues.

5.2.6 Extent of tunnel gullies

Tunnel gully erosion occurs on susceptible soils and lithology throughout various parts of New Zealand. Going by the NZLRI data, tunnel gully erosion is more extensive in the North Island, and less extensive but more severe in the South Island.

Tunnel gully. Photo: *M. Tuohy.*

Gully erosion is found in the North Island on the strongly weathered sandstone downland and hill country of Northland and the colluvial slopes of the tertiary hill country. Tunnel gullies have been recorded in Northland, Coromandel, Waikato Basin Central Volcanic Plateau, Manawatu-Wanganui hill country, Hawkes Bay and the Wairarapa. It occurs on the following rock types:

- In soils derived from strongly weathered sedimentary rocks (humid climate)
- On tephra deposits (cool humid climate)
- On colluvial footslope deposits on sedimentary hill country (humid climate)

In the South Island, tunnel gully erosion occurs mainly on loess hill country soils, and has been noted in Northeast Marlborough, coastal North Canterbury, Banks Peninsula, South Canterbury/North Otago Downlands, Otago Peninsula and Mid-Clutha Valley. It occurs on loess-mantled slopes and mixed loess/colluvium slopes with yellow-grey earth soil groups (pallic soils) in sub-humid to semi-arid climate.

5.2.7 *Mountain gullies*

These form where debris avalanche scars and debris flow deposits are eroded by runoff. They are found in the greywacke and schist mountain lands of both islands; and are present in low forested ranges (below 1000 metres a.s.l.), as well as the alpine tussock lands.

Chapter 6, Mountain Lands Erosion, describes mountain gullies, noting that they are a natural phenomenon and nearly impossible to control.

The NZLRI map does not show deep-seated gullies' location separately from the other forms. They are concentrated in distinct regions, and on distinct geological formations:

- soft marine sediments: sand of the Kaipara and Manukau barriers; sands and silts of the dissected terrace country extending from Hawera to Pohangina; pockets of sandy gravel marginal to the hills of Wairarapa, Marlborough and North Canterbury; dissected glaciofluvial gravels in inland Canterbury and Otago
- cemented marine sediments: mudstone formations containing swelling clay, throughout Gisborne, Hawkes Bay and Wairarapa; also small pockets in Marlborough and North Canterbury.
- crushed rocks: argillites of late Cretaceous and early Tertiary age, marginal to the Raukumara Range (Gisborne), Urewera Range (northern Hawkes Bay) and Haurangi Range (eastern Wairarapa).

5.3 PRINCIPLES OF CONTROL

5.3.1 *U-shaped gullies*

- Control stormwater runoff over the gully head
- Control stormwater runoff through the gully floor
- Stabilise the gully head
- Stabilise the gully floor
- Stabilise the gully sides.

The following practices are used in controlling U-shaped gullies:

- Control of runoff over gully head or away from the gully head. There is a wide range of options including diversions, flumes, pipes and drop structures.

- Reduction in peak runoff rates in combination with stabilising gully head and controlling runoff over gully head.
- Planting of gully head (small gullies only)
- Planting of strong points at critical locations.
- Retirement of gullies in association with runoff control.
- Ground contouring to "smooth out" small gullies on low terraces (in combination with runoff control and surface vegetation).

Where the soils or geology is susceptible to u shaped gully erosion, the following matters should be considered to ensure that gully erosion is avoided:

- Avoid exposure of bare ground, particularly on overland flow paths
- Avoid concentration of stormwater runoff as far as practicable
- Avoid digging drains to discharge pounded water where the soils are non-cohesive and susceptible to fluvial erosion, and the discharge point is uncontrolled
- Avoid uncontrolled discharge of stormwater over steep drop-offs
- Reduce peak flood flows within catchments by using flood detention structures and vegetation as far as practicable
- Control stormwater runoff so that overland flow paths are well stabilised

5.3.2 Tunnel gullies

- Divert and control overland flow to controlled surface outlets
- Retain a dense ground cover on overland flow paths
- Stabilise by planting poles directly in collapse holes, and also in between them.

When tunnel gully erosion problems are only slight to moderate, and occur as holes or depressions, remedial works can be undertaken by pole planting using oversized willow poles planted directly in the tunnel gully hole. Where the erosion feature has formed collapsed gullies, pair pole or staggered pole planting can be carried out. In some cases, seedling trees within protectors may be used as an alternative to poles.

In more severe situations where the tunnel gullies become totally collapsed and exposed, and the gully erosion continues to worsen, more extensive works are required to remedy the problem. In these situations, works include diversion and control of overland flow to controlled surface outlets, and revegetation of dense stable ground cover. Contouring to infill gullies in association with runoff controls may also be an option. Other revegetation options are also likely to be required. These include tree planting (including space planting or afforestation) and temporary or permanent retirement of the affected area from grazing.

Attempting to control tunnel gullies by ripping and compacting the tunnels, will only provide temporary relief, as the soils are inherently susceptible to tunnel gullying, and this method of treatment only addresses the symptoms of the problem, not the cause.

Preventative works to control tunnel gully erosion are normally only employed when there is evidence of tunnel gully erosion occurring. In these situations, space planting of willow or poplar poles, or protected seedling trees are common methods used to address the problem. Planting can be carried out following the line of the gully erosion if it is evident. In Northland, coral tree (Erythrina spp.) poles can be planted as an alternative to willow poles.

In areas where the soils are known to be susceptible to tunnel gully erosion, normal sustainable farming practices such as maintaining a dense pasture sward, controlling animal and plant pests, minimising bare ground etc should be used to reduce the risk of tunnel gully erosion occurring.

5.3.3 The key principles to control V shaped gullies include

- Control stormwater runoff from the surrounding catchment
- Maintain negative pore water pressure in soil and rock, around the gully's head and sides (de-watering)
- Increase shear strength in surrounding soil and rock (tree root reinforcement)
- Stabilise the gully floor (debris dams or live dams).

Control of V-shaped gullies and the associated erosion that normally occurs with the gully erosion, is largely dependent on control of the eroding channel floor. Once control of this has been achieved, then other works to control erosion on the sides of the V-shaped gully can be used. This often requires using a number of practices in conjunction with each other, or in succession. The main approaches, in order of importance, are stabilising the gully floor, stabilising the sides of the gully system or reducing the peak runoff through the gully system

The Ministry of Works have successfully demonstrated at Kaitangata Station (Gisborne District) that gully control can be achieved by using runoff control as the initial control measure, in conjunction with a range of other practices (Hall 1970).

In almost all situations, long-term control will rely on stabilisation of the unstable land mass. This is generally achieved through some form of tree planting. There are a number of planting options that may be used. Tree planting options include pole planting of the immediate gully toe slopes using poplar or willow pole material, wide space planting of trees on the gully and adjacent slopes with grazing underneath, or afforestation over a reasonable proportion of the gully catchment.

Because V-shaped gullies normally occur in conjunction with other types of erosion, preventative works are aimed to address a range of erosion forms on the land susceptible to gullying.

Many of the V-shaped gullies have been formed following the removal of indigenous vegetation on steep land and the conversion of the land to pasture. A pastoral land use may not be sustainable without intensive soil conservation planting, and often some form of afforestation may be the only sustainable land use option to avoid or mitigate erosion problems.

When these susceptible areas are afforested, the tree roots help bind the land, and evapotranspiration from the trees pumps water from the soil to reduce the volume and rate of stormwater runoff. A dense vegetative ground cover (particularly on the gully floor) helps to ensure a stable channel, which also assists in reducing the risk of gully erosion.

Therefore, the key to prevention or stabilisation of V shaped gullies is the establishment of vegetation, particularly trees.

5.3.4 Others

Stabilisation of debris avalanche/flow gullies requires engineering measures. Their cost is high, and can only be justified where roads or railways pass through the ranges. Even here, such measures are rarely attempted. Elsewhere in the mountains, the gullies are simply left to revert; a process which will take decades if not hundreds of years.

Ground contouring is sometimes used in conjunction with runoff control and surface revegetation on U-shaped gullies and tunnel gullies. While the costs of recontouring can be high, the land can be brought back into a more productive use. The contouring operations also provide an opportunity to direct stormwater runoff to a safe controlled outlet.

5.4 APPROPRIATE CONTROL PRACTICES

Table 5.1 Control techniques for different gully forms.

Gully Types	Control Principles
U Shaped Gullies	Runoff Control Systems Tree & Pole Planting Surface Protection: Vegetative and Non-Vegetative Mechanical Infilling or Contouring
V Shaped Gullies	Runoff Control Systems Structures to Stabilise Gully Head/Floor Tree & Pole Planting Surface Protection: Vegetative and Non-Vegetative
Tunnel Gullies	Runoff Control Systems
	Tree & Pole Planting Surface Protection: Vegetative and Non-Vegetative
	Mechanical Infilling or Contouring

Descriptions of the various practices for control of deep seated Gully Erosion are found in chapters/sections of Part B:	
3.2	Fencing Management for Erosion Control - Hill Country
3.3	Fencing Management for Erosion Control - Mountain Land
8.2. (8.3)	Managed Reversion of Retired Land - Hill Country
11	Gully Control Structures
12	Pole Planting
13	Erosion Control Forestry

5.5 REFERENCES

Bartleet JR 1965: The Role of Contouring in the Conservation Programme of North Auckland. Soil conservation paper from the Conservation Subsection of the 11th New Zealand Science Congress, Auckland, 11-17 February 1965.

Bates RL; Jackson JA 1980: Glossary of Geology 2nd Edition. American Geological Institute, USA.

Bennett HH 1947: Elements of Soil Conservation 1st Edition. New York and London, McGraw-Hill Book Company Inc.

Borlase OM 1967: Gully Control by Check Dams – Poverty Bay Catchment Board Report No.1805, First Revision 1967.

Brice JC 1966: Erosion and deposition in the loess-mantled Great Plains, Medicine Creek Drainage Basin, Nebraska. Geological Survey Professional Paper 352H. United States Department of the Interior.

Cathcart RW 1998: Soil Conservation Fact Sheet 2; Gully Control in Northland. Unpublished fact sheet from Northland Regional Council.

De Rose, R. et al 1998: Gully erosion in Mangatu Forest, New Zealand, estimated from digital elevation models. Earth Surface Processes and Landforms, v23, pp1045-1053.

Environment BOP 1998: Sustainable Options Fact Sheet SC10/98. Detention Dams and Drop Structures.

Eyles GO 1983: The distribution and severity of present soil erosion in New Zealand. New Zealand Geographer 39(1):12-28.

Eyles GO 1985: The New Zealand Land Resource Inventory Erosion Classification. Miscellaneous Water & Soil Publication Number 85, National Water & Soil Conservation Authority

Eyles GO 1993: Gully erosion control techniques for pumice lands. Manaaki Whenua Landcare Research New Zealand Limited, Lincoln New Zealand in association with Environment Waikato.

Eyles GO: Towards a New Zealand Erosion Form Classification. Unpublished notes on types of gully erosion.

Gage, M. and Black, R. 1979: Slope stability investigations at Mangatu State Forest. Technical Publication 66, FRI.

Gair J 1973: Pole and netting dams. A paper presented as part of a seminar "Prevention and Control of Mass Movement and Gully Erosion", arranged by Ministry of Works in conjunction with New Zealand Association of Soil Conservators and Poverty Bay Catchment Board, 15-19 April 1973.

Hall, J. de Z. 1986: Water control works, Gisborne-East Cape region, 19691979. Unpublished report, Water and Soil Division, MWD.

Howie WR 1968: A Design Procedure for Gully Control by Check Dams, Water and Soil Division, Ministry of Works, New Zealand May 1968.

Humes Industries Limited: Drainage Products Technical Manual

Leopold, L.B., Wolman, M.G. and Miller, J.P. 1964: Fluvial processes in geomorphology. 522pp, Freeman & Co, San Francisco.

Lynn IH; Eyles GO 1984: Distribution and severity of tunnel gully erosion in New Zealand. New Zealand Journal of Science Vol.27 175-186.

McCaskill LW 1973: Hold This Land, A History of Soil Conservation in New Zealand. Published by AH & AW Reed, Wellington, Sydney, London.

NWASCO 1973: Prevention and Control of Mass Movement and Gully Erosion. Proceedings of a seminar arranged by Ministry of Works in conjunction with New Zealand Association of Soil Conservators and Poverty Bay Catchment Board, 15-19 April 1973.

NWASCO 1975-79: 'New Zealand Land Resource Inventory Worksheets', 1:63,360. National Water and Soil Conservation Organisation, Wellington, New Zealand.

Phillips, C., Marden, M. and Miller, D.E. 2000: Review of plant performance for erosion control in the east Coast Region. Contract Report LC9900/111 to MAF, Landcare Research.

Poverty Bay Catchment Board 1978: Report of Land Use Planning and Development Study for Erosion Prone Land of the East Cape Region: Section 1 The East Coast (Red Report), May 1978.

Wilkie DR: Restoration of Hill Country by Conservation Farming, Wither Reserve, Blenheim. Bulletin No.8, Soil Conservation and Rivers Control Council.

Chapter 6

Mountain Lands Erosion

6.1 INTRODUCTION

Mountain land erosion includes needle ice (frost heave), sheet wash, wind blow, periglacial creep of soil and rock fragments, together with erosion of gullies by debris avalanches, debris flows and torrents.

Natural or geological erosion has been defined as the gradual wearing away of the land by water, wind and gravity. It is sometimes called "normal" erosion but this term is open to criticism on the grounds that in nature normality is difficult to define. In mountain lands natural erosion becomes increasingly prevalent at higher altitudes.

"Induced" (or "accelerated") erosion results from the disturbance of the environment by human activities. ("Induced" is the better term.) For management, it is important to distinguish between natural and induced, but there are many situations where there is no clear answer as to whether the erosion is induced or part of the natural cycle.

Spectacular mountain land erosion is not a recent development in New Zealand. Strange (in 1850) and Colenso (in 1884) climbed to various high points on both islands and described various landslides and severe erosion.

The short tussock grasslands of the high country are a recent vegetation type. Three thousand years ago the high country of both island was a mosaic of forest, scrub and tall tussock. Short tussock was present as a successional plant community on small areas burnt by natural fires, such as those resulting from lightning strike or volcanic eruption. There was a sudden increase in fire frequency 900 years ago when the Maori arrived, and this continued, in particular, when Europeans arrived. Because of the fires, much of the lowland and montane forests was replaced by short tussock grassland.

In the 1850s and 1860s, settlers brought herbivores such as sheep, cattle, deer, chamois, thar, goats and rabbits to this grassland. Regular fires continued as part of the European land management practice and although regulated by 1913 legislation, periodic burning continued unabated until the Soil Conservation and River Control Act in 1941 enforced fire control. From the 1870s through the 1940s, rabbits proliferated through the tussock. Too-frequent burning, followed by too-heavy stocking, followed by rabbits, depleted the short tussock grasslands; and to a lesser extent, the high-altitude tall tussock. Topsoil was exposed to surface erosion by ice, water and wind.

6.1.1 History of research

One of the earliest accounts of soil erosion in the high country (Gibbs and Raeside, 1945) showed that a quarter of the land was extremely eroded, with less than 50 percent of the topsoil remaining, and that only 20 percent of the area had slight or no accelerated erosion. Gibbs and Raeside calculated that the surveyed soil depletion amounted to a total loss of about 1.5 billion tonnes of soil. Research in the last four decades has increased our understanding of the types, magnitude, frequency and triggers of erosion in the New Zealand mountain lands. Much of the work, however, has been into the larger discrete forms like debris avalanches and mountain torrents, rather than the more extensive forms produced by frost heave and wind (Whitehouse 1984).

Screes and gullies on grazed tussock rangeland, Torlesse Range, Canterbury.
Photo: D Hicks.

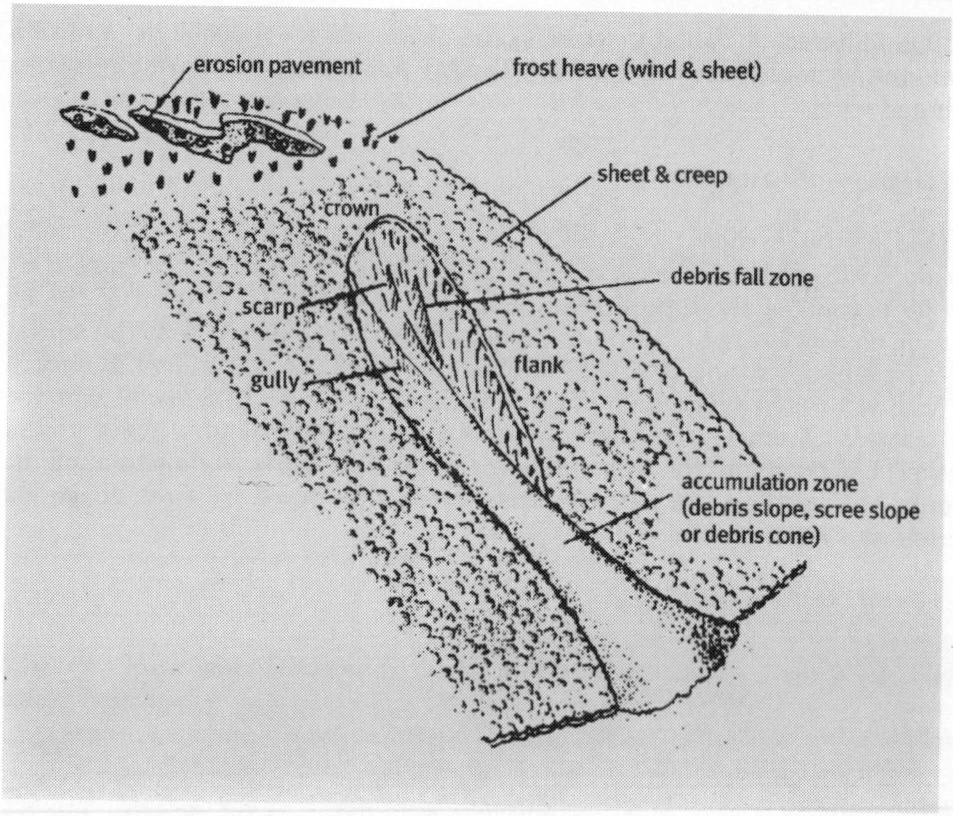

Fig 6.1 Diagram of common forms of mountain land erosion in New Zealand
(after Cunningham, 1978)

6.2 TYPES, PROCESSES AND EXTENT OF EROSION IN THE MOUNTAINS

6.2.1 Frost heave (needle ice formation)

Needle ice formation will be defined here as a separate erosion process which enables particles to be removed later by sheet and wind erosion process. The needle ice process is more severe where bare ground exists and is found in all mountain areas especially higher than 900 m above sea level. Ice pulverises the top of exposed soils and on drying the material is easily removed by wind or rain.

Gradwell (1960) and Soons and Greenland (1970) observed needle ice and noted that the growth of ice needles depends mainly on the rate of heat loss and the availability of water. Hayward and Barton (1969) used a movie film to record the disruptive action of needle ice and showed 22 freeze-thaw sequences moving surface material up to one metre down slope. Subsequent advances in our knowledge of this slow erosion, operating in the Southern Alps have not been extensive (Whitehouse, 1984).

O`Loughlin (1984) reports slopewash rates of 150 t/km²/yr measured on bare subsoil surfaces exposed to frost heave in Hut Creek, Craigieburn Range. Following planting of Pinus contorta on the site, soil loss declined rapidly from tree age 4, until at tree age 10 soil loss was negligible.

Altitude and climatic zonation plays a significant role in the major process of needle ice formation where we find frost heave occurring in the semi arid inland basins of Otago and Canterbury from as low as 300-350 m above sea level, particularly where winter frosts and permafrost occur. Frost heave commonly occurs near Alexandra and the intermountain basins of Canterbury – that is, the Mackenzie Basin, where altitude can vary from 350 m (flats and terraces) to 1500 m, or to the tops of the mountains.

6.2.2 Sheet erosion

Sheet erosion is the removal of a thin surface layer of topsoil or subsoil by a layer of water flowing over the land. The impact of raindrops on the exposed soil breaks up the surface layer and loosens particles, which can then be moved. If it continues for some time an infertile stone pavement results.

Soil, defined as particles less than 2 mm in diameter in Hayward's study (1969), appears to be eroded from surfaces covered with rock particles (scree and erosion pavement) at about (0.012 +/-0.004 kgm²/yr. This is some four times faster than from totally vegetated sites (0.0028 + 0.0005 kg/m²/yr and some three times less than from non-vegetated surfaces other than scree.

6.2.3 Wind erosion

In the mountain lands wind erosion is widespread on exposed tops, sunny faces, and areas that have had the vegetative cover denuded. It is possibly more pronounced in the South Island due to the persistent and often prolonged North West foehn winds.

Wind moves soil particles by the processes of deflation and abrasion. (The chapter on wind erosion gives a fuller description of these processes.) The amount of soil in transit in the wind has been measured at six sites in the Rakaia, Waimakariri and Ashburton Basins by Butterfield (1971), using 7-cm diameter traps located at various heights above the ground. Over the year of measurement, wind erosion was greatest in early spring to early summer with the lowest levels of erosion occurring during mid summer (January) and in winter. Because of the difficulty, wind erosion rates have not been measured in the high country since Butterfield's study.

Recent research work using caesium-137 (Hewitt, 1996) into surface soil losses in Central Otago on the lower foothills and terraces near Alexandra (300–320 m above sea level) has shown that a sunny side slope lost 50 percent of its expected caesium-137 input through erosion of top soils. This was equivalent to a loss of 3.4cm of soil over 40 years. Shady side slopes and foot slope sites gained caesium-137 through net soil deposition mainly from the northwest winds. This information indicates that the surface soil losses are not as severe as first thought (even though bare ground may be significant at certain times, such as after a drought or rabbit infestation).

Frost heave, sheet erosion and wind erosion are the only mountain land erosion processes for which soil conservation would normally be attempted. While natural, they have been accelerated at low to middle altitudes by 150 years of vegetation disturbance – burning of tussock grassland, followed by sheep grazing, followed by rabbit infestation, followed by weed colonisation. In these zones, the erosion can be reduced by avoidance or restriction of burning, revegetation of depleted tussock, lenient grazing management, and rigorous control of pests and weeds.

Other erosion processes occur on alpine slopes and ridges, where soil conservation would not normally be attempted. Techniques for their control are beyond the scope of this handbook. They are described here just to remind readers that these processes are encountered in the mountains; are natural; and are difficult to control.

6.2.4 Periglacial erosion

At high altitudes, soil in the mountains is extensively disturbed by periglacial erosion processes. Frost heave, already described, is one of these. Repeated freeze-thaw cycles sort the skeletal soil into "patterned ground", with stone rings or stone stripes separating patches of silty or sandy soil.

Soil creep is a slow, plastic flow of surficial soil and rock debris downslope, creating solifluction terraces or solifluction lobes. At high altitudes it is driven by changes in stress and resistance, as the soil alternately freezes and thaws. Scree creep is a related phenomenon, occurring on talus slopes beneath bluffs. The talus initially accumulates from rockfall or debris avalanches. Where vegetation can establish, it stabilises; but where it cannot (or if it is breached), the shattered pieces of rock may be slowly shifted downslope by freeze and thaw of water in the cracks between.

6.2.5 Debris avalanches

Debris avalanches are slide-type mass movements of soil or rock on steep mountain slopes. They are triggered by the same mechanism as shallow landslides (soil slips) – see Chapter 4 – and start in a similar position, a hollow which concentrates soil water draining from farther upslope. Once triggered, the mountain slope's steepness ensures that landslide debris, instead of coming to rest on the slope, keeps moving. Slope length enables the debris to build up momentum sufficient to start scouring away vegetation and soil in its path. The result is a long, narrow scar stretching from upper slope to slope foot, with a jumbled deposit of soil, rock and timber at the base. Repeated debris avalanching may build up a talus cone in the lower part of the scar.

6.2.6 Debris flows

A debris avalanche may transform into a debris flow as it descends. This happens if the debris incorporates sufficient water or air in its downward passage to turn it to fluid. The fluidised debris keeps moving across the valley floor instead of coming to rest at the slope foot. If the floor is wide, repeated debris flows will build up an alluvial fan, tapering from and angle of 20° or more at its apex, down to 5 or 6° at its foot where the debris gives out or settles. If the valley floor is narrow, the debris will turn and flow downstream for distances of several hundred metres until it gives out. This will be at a point where channel gradient is about 5 to 6°, and a graded deposit of alluvium interspersed with boulders and timber will be left.

6.2.7 Mountain gullies

A debris avalanche's linear scar is a natural conduit for runoff from the upper mountain slope above and around it. Subsoil is exposed in its sides, and shattered rock in its base. Many such scars turn into gullies, but it is important to remember that such gullies are maintained as much by occasional debris avalanches during storms as by runoff during lesser rainfalls. The same holds true for gullies in the fans and valley-bottom deposits left behind by debris flows.

6.2.8 Torrents

A torrent is a permanently flowing stream going down a mountain or flowing along a mountain valley bottom. It differs from normal streams in that its gradient is steep, typically exceeding 5° and occasionally in excess of 20, interspersed with rapids and waterfalls.

A torrent's steep gradient, combined with frequent passage of short sharp floods, and a sediment load of coarse gravel and small boulders, give it great erosive energy. The bed and banks are eroded in almost every flood; the small, discontinuous pockets of coarse alluvium are rapidly worked downstream, and supplied to larger streams and rivers which exit from the mountains. Where slopes are susceptible to debris avalanches and the avalanches can reach the channels, alpine torrents are also subject to passage of debris flows. These cause even more erosion and deposition than the floods.

The definitive studies of mountain erosion processes are published in Japanese, Chinese, German or Italian. For an English-language summary, refer to Scheidegger (1988). Susceptibility of the New Zealand landscape is summarised in Chapters 6 and 19–22 of Soons and Selby (1992).

6.2.9 Conservation measures to control natural erosion in mountains

Engineering measures cannot prevent the processes described above, but they can control their impact; either stopping or diverting them before they cause damage.

There are no standard practices for the control of debris avalanches, debris flows or mountain torrents in New Zealand as few highways or railways pass through the mountains and control measures are rarely tried. Readers are advised to seek information from one of the countries where such measures are extensively implemented due to settlement and infrastructure in the mountains. Overseas practice appears to be most advanced in Japan, Taiwan, Switzerland, Austria and Italy.

6.2.10 Erosion rates in mountain lands

The most significant aspect of mountain land erosion is the large quantities of silt and material removed from catchments and deposited downstream into lakes onto alluvial flood plains, or transported out to sea. Sediment yields from some basins in the Southern Alps are amongst the highest in the world. Yields vary from about 100 t/km²/yr for basin in the dry intermountain areas (Twizel and Forks Rivers) to about 15,000 t/km²/yr for basins in the western part of the Southern Alps (Hokitika, Cleddau and Haast Rivers) (Hicks D.M. et al 1996).

The most accurate measurement of sediments in Otago comes from Falls Dam, an irrigation reservoir on the upper Manuherikia River. Six hundred thousand tonnes of silt, clay and sand have accumulated behind the dam since 1935. For an assumed reservoir trap-efficiency of 80 percent, this is removal of 44 t/km²/y, from the area of drainage basin above the dam (Bishop et al, 1984). This is a denudation rate of 0.02 mm/yr in the driest part of the mountains. The highest figures, from basins in the western part of the Southern Alps, have a large margin of errors (Hicks et al 1996). Even after allowing for this, they demonstrate the enormity of erosion where rainfall is between 3000 and 10,000 mm a year.

6.3 PRINCIPLES OF CONTROL

Much of the surficial soil erosion is insidious in nature, although debris flows and the like may come only after a major storm. The following factors contribute to erosion in the mountains:

- geology and soil type
- altitude, slope and aspect
- climatic influence related to above microtopography, freeze/thaw regime and growing season
- burning native grassland (especially snow tussock and associated species) and not spelling from stock or a very hot fire
- overgrazing after a burn
- depleted, open ground cover subject to the severe climate and other elements
- low fertility soils with a poor carbon to nitrogen ratio and low organic matter content

- animal pest population increases by rabbits, hares, possums, goats, thar, chamois and feral deer
- plant pest population ingression(eg, gorse, briar, matagouri and *Hieracium* species)

The principles of control for surficial erosion are very similar to that for sheet and rill erosion on the lower plains and downlands but occur in a more diverse climatic environment. Control measures often must be taken at higher altitudes, where plant-growing days are very short.

The primary principle is to manage the land in its present form by maintaining or improving the vegetative protective cover and the soil health status especially the organic matter content. Success in erosion control on these high altitude properties can often only be made by changing land management (such as burning and grazing management) systems over the entire farm unit and not on a paddock (or block) basis.

There has been considerable research work carried out into soil conservation management on the mountain lands in New Zealand by various organisations since the 1941 Soil Conservation Act was passed. This included work by the Forest Research Institute at Craigieburn and in the North Island on the Ruahines, by DSIR Grasslands Division, Ministry of Agriculture, AgResearch, MWD Soil Conservation Centres, Landcare Research, Tussock Grasslands and Mountain Lands Institute, The Hellaby Trust, various universities, catchment authorities (which are now part of the regional councils)are some of these.

6.1 APPROPRIATE CONTROL PRACTICES

Table 6.1 Summary of mountain erosion types and management practice

Type of erosion	Factors to Consider for Management Practice	Examples of Practices (Not inclusive)
Frost heave*	Type and condition of present vegetation	Retirement fencing
Sheet*	Soil type (% topsoil/subsoil left & % organic matter)	Erosion control fence
		Cattle-proofing
	Geology	Rotational grazing
Wind*	Rainfall, storm events	Summer grazing
	Aspect (sunny versus shady)	Aerial oversowing and top dressing
Soil creep**	Altitude & slope	
	Timing	Mulching
	Past management (burning, grazing)	Erosion control forestry
	Plant & animal pests	Protection forestry
	Land Use Capability class & land use	Retirement planting

	Tenure & ownership	Feed banks
	(leasehold, freehold, Crown reserve)	Feral animal control
	Access	Destocking and spelling
	Economics of practice	Burning management
Streambank***	As above plus:	Above as applicable, plus:
Scree*	• size of catchment	• water diversion
Gully***	• type of regolith	• mulching
Debris available**	• type of vegetation required	• toe stabilisation
	• erosion rate and ease of repair	• engineering structures
	• spring dewatering	

* = surface erosion, ** = mass movement, *** = fluvial.
The geomorphology of the types of erosion above are mostly schist, greywacke and volcanic ash.

Descriptions of the various practices for control of Mountain Land Erosion are found in chapters/sections of Part B:

2.3	Tussock Management for Surface Erosion Control
3.2 (3.3)	Fencing Management for Surface Erosion Control - Hill Country and Mountain Land
4.3	Pasture Revegetation - Mountain Lands
6	Burning Management - Mountain Lands
8.4 (8.3)	Managed Reversion of Retired Land - Mountain Lands
13.5	Forest Management Practices - Mountain Lands

6.5 REFERENCES

Ackroyd, P., 1978: Geological history and erosion, Tussock Grasslands and Mountain Lands Institute Review no

Bishop, D.G., et al, 1984: Modern sedimentation at Falls Dam, upper Manuherikia River, Central Otago, New Zealand. New Zealand Journal of Geology and Geophysics 27, 305-312

Butterfield, G. R., 1971: The susceptibility of high country soils to erosion by wind Proceedings NZ. Association Soil Conservators Seminar on Catchment Control, Massey University: 329-345

Colenso, W., 1884: In Memorium. An account of visits to and crossings over, the Ruahine mountain range, New Zealand, and of the natural history of that region. Daily Telegraph, Napier

Cunningham, A., 1978: Erosion Assessment in New Zealand mountains, Proceedings of Conference on Erosion Assessment and Control in New Zealand, published by the New Zealand Association of Soil Conservators

Floate, M., 1991: Guide to Tussock grassland farming booklet Agresearch, Invermay, Mosgiel

Gibbs, H.S., Raeside, J.D. et al 1945: Soil erosion in the high country of the South Island, Dept. of Scientific & Industrial Res. Bulletin, no 92

Gradwell, M.W., 1960: Soil frost action in snow tussock grassland, NZ Journal. Science 3: 580-590

Hayward, J. A, & Barton, J. H., 1969a: Erosion by frost heaving, Soil and Water 6(1): 3-5

Hayward, J. A., 1975: The Torlesse Research Area, Tussock Grasslands and Mountain Lands Institute Review no 31

Hayward, J. A., 1969b: The use of fractional acre plots to predict soil loss from a mountain Catchment. Lincoln Papers in Water Resources 7, New Zealand Agricultural Engineering Institute, Lincoln College 93 p

Hayward, J. A., 1980: Hydrology and stream sediments in a mountain catchment. Tussock Grasslands and Mountain Lands Institute Special Publication 17

Hewitt, A.E., 1996: Estimating surface erosion using caesium 137 at a semi arid site in Central Otago, New Zealand, Journal of the Royal Society of New Zealand Vol 26, no 1, pp 107118.

Hicks, D.M. et al 1996: Variation of suspended sediment yields around New Zealand: the relative importance of rainfall and geology. In Proceedings of Symposium on Erosion and Sediment Yield, Exeter IASH Publication 236.

O'Loughlin, C.L.,1984: Effectiveness of introduced forest vegetation for protection against Landslides and erosion in New Zealand steeplands, in Symposium on Effects of Forest Land use on Erosion and Slope Stability, Honolulu, Hawaii pp 275-280

Owens, I.F., 1967: Mass movement in the Chilton Valley, M.A. Thesis, University Canterbury Library, 92pp

Pearce, A.J., and O`Loughlin, C.L., 1978: Hydrological and geomorphic processes on Forested lands – a selective overview. pp62-76 in; Proceedings of a Conference on Erosion Assessment and Control in New Zealand. New Zealand Association of Soil Conservators

Scheidegger, A. E. 1988: Hazards from mass movements in mountain regions. Pp 21-41 in El Sabh, M. & Murty, T. (eds), Natural and man-made hazards. Reidel Publishing Co.

Soons, J.M., Greenland, D.E., 1970: Observations on the growth of needle ice, Journal Water Resources Research, 6(2):579-5

Soons, J. M. and Selby, M. J. 1992: Landforms of New Zealand. 531pp, Longman Paul, Auckland.

Strange, F., 1850: The Canterbury Plains, in Canterbury papers no 3 London pp77-80

Whitehouse, I.E., 1984: Erosion in the eastern South Island high country- a changing perspective, Tussock Grasslands and Mountain Lands Institute Review no 42.

Chapter 7

Sand Dune Erosion

7.1 INTRODUCTION

Sand dune ecosystems are very fragile ecosystems, which persist under extremely difficult conditions, and their management has been a challenge to New Zealand land managers for a very long time. Dunes are affected by:

- wind, salt, and generally low temperatures from cold sea breezes
- the substrate of mobile sand, which combined with wind, can "sandblast" seedlings and mature plants, and can bury or expose established plants
- temperatures, whose fluctuations can be very large, with frosts in depressions or basins during winter but which can rise rapidly during calm sunny days;
- amount of moisture for plant growth, which is often small because the moisture-holding capacity of sand is limited
- summer droughts that can be quite severe
- nitrogen, which is the most limiting element because it is easily leached out.

7.1.1 History of sand erosion control

On conventional sand dunes, the following stages of reclamation/stabilisation are recognised:

1. establishment and protection of a stable foredune parallel with the coast (see section "Foredune").
2. fixation of the unstable sand area behind the foredune and further inland with binding plants (see section "Main dune")
3. establishment of a permanent vegetative cover using grasses and legumes for pasture development, or trees for afforestation, or native plants for the restoration of reserves and other high value areas (see section "Main Dune").

For details on the first three stages, see van Kraayenoord and Hathaway (1986a) (pages 69-77), and more recent findings, described in the following sections of this chapter. Reference should also be made to the very recent Auckland Regional Council publication, The Coastal Erosion Management Manual.

As a primary sand stabiliser, marram grass (Ammophila arenaria) has been given most prominence. Although it is an introduced plant, it has proven itself as a sand dune stabiliser throughout New Zealand. Since the early 1920s, marram grass has been planted, more recently with mechanical planters, to stabilise sand dunes because it is easy to propagate and establish.

The New Zealand silvery sand grass (Spinifex sericeus, previously hirsutus) is also a true sand binder, found on fore dunes (Whitehead, 1964), but it is not easily propagated (Sale, 1985). Through the voluntary assistance provided by Beach Care and Coast Care groups, spinifex has in recent years gained more prominence in dune revegetation programmes. Detailed propagation methods may be found in regional council leaflets, such as Bay of Plenty Regional Council Information Brochure no. 6. (See also Bergin, 1999).

The other New Zealand sand binder, pingao (Desmoschoenus spiralis), is usually found a little further inland where under favourable conditions it forms rather irregular hillocks (Whitehead, 1964; van Kraayenoord and Hathaway, 1986a). Hillocks can lead to wind funnelling and thus to localised erosion, which is to be avoided. Pingao is more difficult to propagate than marram grass, but Herbert and Bergin showed that it could be used in duneland rehabilitation (see Bergin and Herbert, 1998). Pingao is highly sought after in Maori weaving due to its yellow colour.

7.1.2 The "natural character" of sand dunes

The preference or desirability of using native plants versus introduced plants for sand dune stabilisation to maintain the natural character of dunes continues to be the subject of much debate (Gadgil and Ede, 1998). In their comprehensive review of sand dune stabilisation research and practice, they say, "Greater success, especially in emergency situations and remote areas, is likely to be achieved by preliminary use of exotic species to fix nitrogen and establish continuous cover, followed by gradual enrichment or replacement with native plants. This approach would satisfy requirements for both erosion control and a return to the natural character of the dunes."

7.1.3 Preventive works

Preventive works focus on avoidance of activities that may disturb the fragile vegetation on sand dunes. Where blowouts have occurred, steps must be taken immediately to remedy this, so as to avoid the problem escalating rapidly.

Human activities, such as walking, can often be channelled onto stable areas formed by walkways and board walks. But 4-wheel drive and other motorised activities are less easily controlled, and barriers may have to be put in place. Building activities in dune areas should be prohibited where possible; otherwise they should be strictly regulated in Coastal Hazard Strategy documents and supervised.

Animal grazing and browsing should be strictly controlled if it cannot be prohibited altogether. Domestic animals can usually be excluded from dunelands with the co-operation of land occupiers. Feral animals should be tightly controlled.

7.1.4 The Coastal Dune Vegetation Network

The Coastal Dune Vegetation Network (CDVN) was formed in 1997. It provides linkages between a wide range of agencies, interest groups, iwi, nurseries, and consultants having a mutual concern for the rehabilitation of degraded sand dunes, particularly revegetation techniques incorporating indigenous coastal species. Financial members include Regional and District Councils, forest companies owning sand dune forests, and the Department of Conservation.

For more information, contact the CDVN secretary, Forest Research, Private Bag 3020, Rotorua. Phone (07) 347 5899; fax (07) 347 5332.

7.2 FORMS, PROCESSES, EXTENT

Sand dune ecosystems are regarded as very fragile because they persist under extremely difficult conditions. Dunes are the most dynamic ecosystems in our environment; human activities or those of animals very easily disturb them. They require constant vigilance and maintenance if they are not to affect various assets. The erosion process in sand dunes is primarily driven by wind that lifts, transports and deposits sand grains into mobile, unvegetated accumulations of sand, called dunes. The dunes may become vegetated and stabilised by highly specialised pioneering plants that tolerate the sandblasting effect of the wind.

When discussing sand dune stabilisation, three types of dune are recognised: foredunes, main dunes, and dunes derived from sandy cliffs. All dunes have been formed by wind action on sand grains, derived from eroded rocks. At the beginning, wind lifts, transports and deposits sand grains into mobile, unvegetated accumulations; these may then become stabilised by highly specialised pioneering plants. The stabilised dunes can become a single ridge along the coast, or dune systems extending many kilometres inland.

The **foredune** (or frontal dune) immediately faces the sea and is the most recent sand dune in a sandy or gravelly beach system. It is the ridge of sand immediately above the high tide mark and may be covered by pioneering plant species such as spinifex, pingao or marram.

The sandy foredune is part of the beach itself and, along with the in-shore seabed, this system forms a natural buffer between the sea and the land. The beach and its associated dune system are able to absorb the energy of the waves, unlike a hard rocky shore, which both reflects the waves and is eroded by them. This mobile beach is able to absorb wave energy because of its gentle slope and the mobile and homogeneous texture of the sand. It protects the area behind the fore dune.

The **main dune** (or back dune) is the sandy area behind the unstable fore dune. The main dune is usually well vegetated and thus does not require stabilisation. However, inappropriate management can lead to local areas of instability which then do require stabilisation, as for fore dunes. In many parts of New Zealand, pasture or introduced forest trees have replaced the original sand dune vegetation.

Sandy cliffs that consist of indurated sand can erode to form sand dunes. Streams often dissect the cliffs so that valleys have been formed, and these can act as funnels for coastal winds. High-velocity winds can abrade the sandy cliffs to form moving dunes that can invade agricultural lands, local reserves, or other assets.

The total area of coastal sand dunes in New Zealand is 305,000 ha, according to NZLRI Data. Of this, 240,000 ha are in the North Island, especially along the western coasts of Northland, Waikato, Rangitikei and the Manawatu; they range from 3- 20 km in width (van Kraayenoord, 1986a). Coastal sand dunes in the South Island are scattered throughout Nelson, Marlborough, Canterbury, Otago, and Southland.

Patrick Hesp (2000)states that sand dunes are a distinctive feature of about 1100 km of the New Zealand coastline. The sand is derived from erosion debris transported along the coast by coastal currents. Rivers bring down the debris, but in some coastal locations, the sand is derived from Pleistocene sandy cliffs, as at Awhitu Peninsula. Most sand is predominantly quartz but some may have such a very high mineral content that it is mined (ironsands). Seashells and rock fragments may be found among sand grains, which can range from coarse sand (2.0–0.2 mm diameter) to fine sand (0.2–0.02 mm).

7.3 PRINCIPLES OF CONTROL

The control of sand dune erosion is based on mitigating the effect of wind on sand transportation. This is usually done by assisting common colonising plants with supplementary planting, or by using non-vegetative materials, such as straw/hay bales and various wind-stilling barriers. Barriers, board walks, and fencing can regulate human and animal traffic. Domestic animals rarely have a place in dune management practices, unless very carefully controlled. Animal pests should be strictly controlled in these fragile ecosystems.

During periods when the dune is relatively stable, vegetation may become established spontaneously from creeping plants such as spinifex and pingao, or these plants may be established from nursery grown stock. In large, unstable areas, marram may be planted, often by means of mechanical planters. The most effective planting is done in very early spring when moisture is available and desiccating winds less frequent.

Regular inspections and prompt maintenance should prevent erosion in dune systems, since these ecosystems are very easily disturbed. In particular, the occurrence of blowouts should be prevented where wind excavates the regular contour of the dune by funnelling. This can destabilise not only the foredune but also the main dune and the areas behind it, usually by covering them with unwanted sand. Hesp (2000) describes the process and his figure 2 illustrates this.

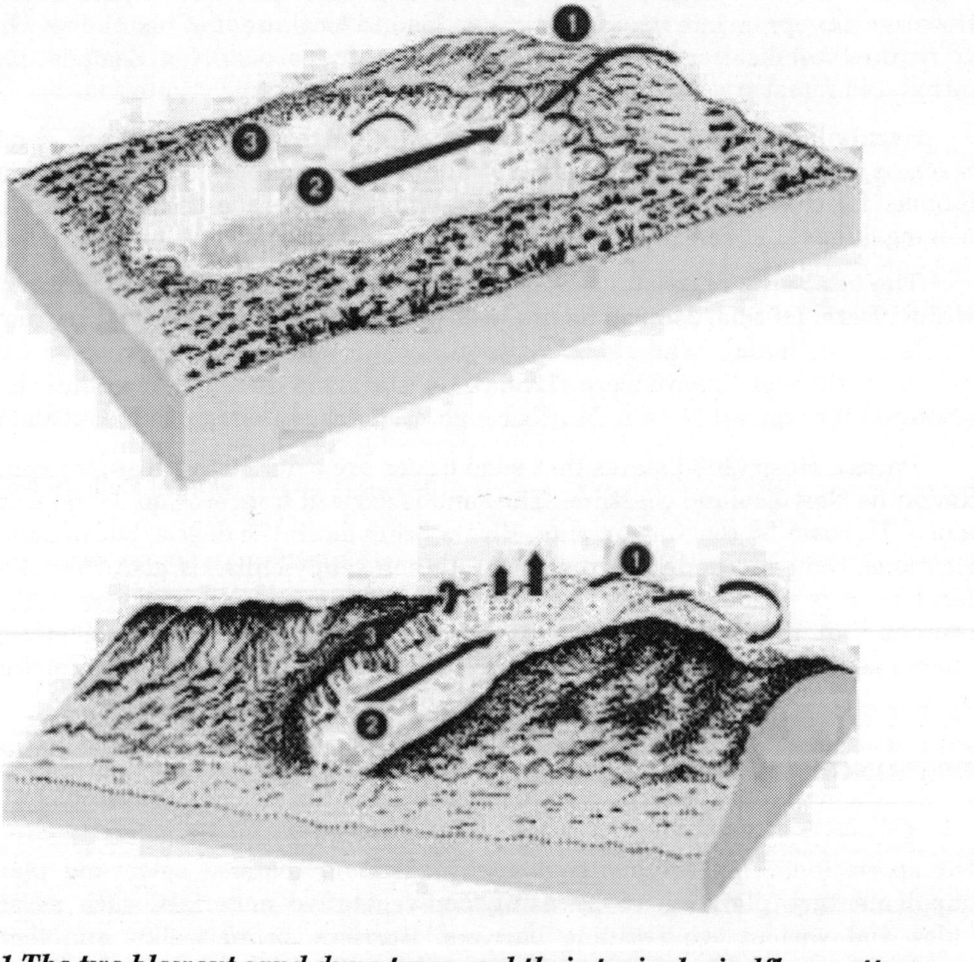

Fig 7.1 The two blowout sand dune types and their typical windflow patterns (P Hesp).
1. Depositional lobe; 2. Deflation basin; 3. Erosional walls

Because of the usually unwanted sand deposition, it is imperative that blowouts be controlled as soon as practicable, irrespective of the time of the year. If planting is unlikely to be effective, non-vegetative means such as straw/hay bales may have to be used, and various wind stilling barriers, such as brushwood fences or sheltercloth may have to be erected. Frequently, both vegetative and non-vegetative materials have to be used at the same time. In some situations, a reshaping of the contours may have to precede the erection of barriers or the planting of vegetation, if the funnelling effect of the wind may destroy any reinstatement efforts.

7.4 APPROPRIATE CONTROL PRACTICES

Descriptions of the various practices for controlling sand dune erosion found in chapters/section of Part B:	
5	Sand Dune Stabilisation Practices
15	Shelterbelts

7.5 BIBLIOGRAPHY

The publications from the Coastal Dune Vegetation Network (CDVN) are essential reading for sand dune managers, as well as the Auckland Regional Council's Coastal Erosion Management Manual. Various regional councils have already produced their own locally adapted bulletins and leaflets; a few examples are given.

CDVN Technical Bulletin Series

Bergin, D O & Herbert, J W 1998, Pingao on Coastal Sand Dunes. CDVN Technical Bulletin No. 1. Publ. New Zealand Forest Research Institute Ltd, Rotorua. 20 pages

Bergin, David 1999, Spinifex on Coastal Sand Dunes. CDVN Technical Bulletin No. 2. Publ. New Zealand Forest Research Institute Ltd, Rotorua. 32 pages

Bergin, David 2000, Sand Tussock on Coastal Sand Dunes. CDVN Technical Bulletin No. 3. Publ. New Zealand Forest Research Institute Ltd, Rotorua. 16 pages.

Hesp, Patrick A 2000, Coastal Sand Dunes – Form and Function. CDVN Technical Bulletin No. 4. Publ. Forest Research, Rotorua. 28 pages.

Auckland Regional Council

The "Coastal Erosion Management Manual", available from ARC in hard copy format at $129.00 incl. GST, or as a CD-ROM at $49.00 in MWS format. Phone (09) 3662000.

Other New Zealand publications

Gadgil, R L & Ede, F J 1998, Application of scientific principles to sand dune stabilisation in New Zealand: past progress and future needs. Land Degradation & Development 9, 131

Pearse, H C H 1958, Stabilisation of sand drifts with blue lupins. NZ Journal of Agriculture 97 (4), 315-324

Sale, E V 1985(?), Forest of Sand – The Story of Aupouri State Forest. NZ Forest Service, Wellington.

van Kraayenoord, C W S 1986a, Plant Materials for Sand Dune Stabilisation, pp. 69-77 in: van Kraayenoord, C W S, and Hathaway, R L (Eds.) 1986, Plant Materials Handbook for Soil Conservation, Vol. 1: Principles and Practices. NWASCA Water & Soil Misc. Pub. No. 93.

van Kraayenoord, C W S 1986b, Management and Uses of Ammophila arenaria (Marram Grass) and Spinifex sericeus (Silvery Sand Grass), pp. 2469 in: van Kraayenoord, C W S, and Hathaway, R L (Eds.) 1986, Plant Materials Handbook for Soil Conservation, Vol. 2: Introduced Plants. NWASCA Water & Soil Misc. Pub. No. 94.

Whitehead, P S 1964, Sand dune reclamation in New Zealand. NZ Journal of Forestry 9 (2), 146-153

Australian publications

NSW Public Works Department and Soil Conservation Service 1987 [Reprint 1995], Beach Dunes – their use and management. Publ. NSW Government Printing Office, Ultimo NSW, 32 pages

Soil Conservation Service of NSW 1990 "Coastal Dune Management – A Manual of Coastal Dune Management and Rehabilitation Techniques", Eds.: Conacher, P.A., Joy, D.W.B., Stanley, R.J., and Treffry, P.T. Soil Conservation Service of NSW, Sydney. 76 pages

"NSW Govt. Coastline Management Manual". This is being updated by Dr Rod Kidd of the NSW Dept of Land and Water Conservation. The manual may be found at:

http://www.environment.gov.au/marine/ manuals_reports/coast_manual/index. html

Regional Council and Department of Conservation publications:

Several regional councils have produced a number of leaflets/publications on sand dune rehabilitation and management. This list is not comprehensive:

Environment Waikato

- Dune Care Code
- Spinifex Seed Collection and Propagation
- Pingao Seed Collection and Propagation
- Guidelines for Spinifex Seed-Head Collection
- How Beaches and Dunes work, Coastal Fact Sheet 13
- Why Dunes are important, Coastal Fact Sheet 14
- Care Groups in Action, October 2000
- Waikato Region Beachcare Newsletter, Christmas 1999
- Website: http://www.ew.govt.nz/educationprog rammes/beachcare/

Northland Regional Council

- Guideline for Stabilising Sand Blows
- Guideline for Protecting Vegetation on Cliffs
- Guideline for Pasture Management on Sand Country
- Guideline for Establishing Permanent Tree Cover on Sand Country
- [Note: these guidelines are based on those prepared for the Awhitu Peninsula Land Group, and are also available from the Group and from Environment Waikato]
- Website http://www.nrc.govt.nz/

Bay of Plenty Regional Council

- Coastcare BOP Programme: Coastal Plants, Spinifex. Coast Care Info. Brochure No. 6

- Website : http://www.boprc.govt.nz/search/

Otago Regional Council

- St Clair to Lawyers Head: Sand Dune Management Programme
- Ocean Grove Recreation Reserve: Sand Dune Stabilisation Programme
- Website http://www.orc.govt.nz/

Department of Conservation

- Project Crimson Supporters Kit, see also Website http://www.projectcrimson.org.nz
- NZ Native Plants in Design – Coastal Forests
- Website http://www.doc.govt.nz

Chapter 8

Earthworks

8.1 INTRODUCTION

Earthworks can induce major, but usually short-lived, changes in sediment yield. American information suggests that that the sediment yields from urban developing areas can be extremely high, sometimes reaching values of 50,000 t/km²/yr (Novotny et al 1981). Figures of 10–12,000 t/km²/yr are quoted from areas undergoing construction in New Zealand (Herald, 1989; Williamson, 1991).

Studies in the Auckland region bear this out, with construction sites yielding 10–100 times more sediment than untouched land (ARC 1992a; ARA, ARWB 1983a, Swales 1989). Another relevant study gathered sediment yield data from five different land uses around the Auckland.

Table 8.1 Annual soil loss in the Auckland region Auckland Proposed Regional Plan: Sediment Control, September 1995)

Landuse	Measured (t/km²/yr)	Average Annual Soil Loss* (t/km²/yr)
Pasture	49	46
Market Gardening	49	52
Developed Urban - Industrial	107	100
Developed Urban - Residential	24	24
Earthworks	6,600	16,800

* Predicted over 20 year period

The predicted yields are a statistical estimation of sediment yields averaged over periods longer than the flow record. A 20-year average was used, as it was considered to be acceptably long-term, (Hicks 1994). The confidence limits of the long-term annual yield varied from a factor of two for the urban and pasture information up to a factor of 4-5 for earthworks. The ARC report considered that the larger, more infrequent storm events yielded disproportionately more sediment from earthwork sites than smaller events. In comparison, the very frequent event yielded most sediment from the established urban areas (because of sediment exhaustion with higher storms).

8.1.1 Impact of earthworks erosion on the environment

Earthworks result in elevated levels of sediment in waterbodies and added deposition on their beds and banks. Water clarity and aesthetic appeal is reduced. High suspended solids concentrations damage or kill stream plants and animals. Reduced plant growth flows through the food chain, to reduce insect and fish numbers. Increased channel erosion and remobilisation of sediments may maintain high suspended solids levels for several years. The suitability of water for uses such as irrigation, rural water supplies, stock watering and recreation can be reduced (ARC, 1996).

Watercourses can be in-filled, causing loss of flood capacity. Sediment can be deposited on floodplains, causing crop and pasture losses, property damage, and burial of fences, roads and bridges. Large volumes can be discharged into estuaries and the ocean.

8.2 TYPES, PROCESSES AND EXTENT OF EROSION ON EARTHWORKS

On earthworks, the basic erosion process is detachment, transport and sedimentation. Water is the usual eroding agent and transporting medium through raindrop dislodging exposed soil particles, and overland flow transporting them downslope. Channelised stream runoff also transports the eroded soil particles to the final receiving environment.

On earthwork sites there are four main erosion types:- sheet erosion, rill erosion, gully and channel erosion.

- Sheet erosion: the removal of a fairly uniform of surface soil by water runoff
- Rill erosion: small channels formed by concentrated runoff flows
- Gully erosion: deep channels scoured out by concentrated runoff flows
- Channel erosion: bed and channel banks are removed by flowing water.

Choice of practices will be dictated by the nature of soil at the construction site. Practices appropriate for stabilising sand or gravel will not work for silt or clay (and vice versa). For a discussion of how different particle sizes are detached, transported and deposited, readers are referred to Bagnold (1965).

8.3 PRINCIPLES OF EROSION CONTROL ON EARTHWORKS

Erosion control measures protect the soil surface against rain and runoff. They consist of site management, water management and stabilisation. Sediment control measures capture eroded soil particles onsite. Fine textured particles are not easily retained once mobilised. Erosion control measures are therefore usually far more effective.

A simple soil loss estimation model, such as the Universal Soil Loss Equation which has been modified for construction sites, can be a very useful tool to work out what factors can reduce erosion and sediment yield.

The following principles are effective in reducing soil erosion and particle transport. They form the basis of an Erosion and Sediment Control Plan

- Keep disturbed areas small and time of exposure short. Stage construction.
- Protect disturbed areas against runoff from above the site i.e. install perimeter controls (see section 4.1.2).
- Keep on-site runoff velocities low.
- Progressively stabilise disturbed areas (section 4.1.3).
- Retain sediment on the site (section 4.2).
- Control erosion at source.
- Fit the development to the existing site conditions. Watch steep areas. Retain watercourses.
- Retain existing vegetation if possible
- Inspect and maintain control measures.
- Temporary and permanent control measures are usually quite different – ensure that design for temporary measures is conservative.

Over the last ten years, erosion and sediment control practices have evolved rapidly in some parts of New Zealand. As more has become known about the effects of sediment on receiving environments, so has the demand for effective controls become more acute. Techniques of control have continued to develop as more is discovered about their strengths and weaknesses. Some have changed significantly over the last few years (eg, sediment retention ponds); while others, like the use of haybales, have hardly changed at all. The principles of control remain the same, but ongoing change with control measures can be expected.

The techniques discussed in this section follow basic principles and will, if implemented and maintained correctly, give good results within their design limits. Remember however, that all of these measures can be expected to change over time.

A simplified layout of erosion and sediment control practices is displayed below. Dust is included as it can be a problem from earthwork sites. This layout includes a selection of the most common measures and is far from exhaustive. Every site is unique and a combination of these, and other practices, can be expected. Innovative site thinking should be encouraged provided new measures are based on sound principles.

8.4 APPROPRIATE CONTROL PRACTICES

Table 8.2 Control techniques for erosion resulting from earthworks

Erosion Control	Practice
Site Management	Site planning and project management that recognises and addresses erosion and sediment control considerations. Runoff Control Systems
Water Management	Diversion channels/bunds Check dams
	Contour drains
	Flumes
Stabilisation	Grassing
	Mulching
Sediment Control Practice	Geotextiles
	Sediment retention ponds Sediment retention bunds
	Silt fences/haybales Storm water inlet protection Pumping
Dust Control	Practice
Planning	Dust management plan
Implementation	Water Dust suppressants Surface stabilisation
	Other options

Descriptions of the various practices for controlling earthworks erosion found in chapters/sections of Part B:	
15	Runoff Control Practices for Earthworks
16	Soil Management Techniques on Earthworks
17	Structures for Runoff and Sediment Control on Earthworks
18	Dust Control Measures for Earthworks

8.5 BIBLIOGRAPHY

ARA, 1983a: Guideline – Earthworks Erosion Management. Upper Waitemata Harbour Catchment Study, Auckland Regional Water Board

ARC 1992a: Erosion and Sediment Control Guidelines for Earthworks. Technical Publication No. 2

ARC 1994: Storm sediment yields from basins with various landuses in Auckland area. Technical Publication No. 51.

ARC 1995: Proposed Regional Plan: Sediment Control. September 1995

ARC 1996: The Environmental Impacts of Accelerated Erosion and Sedimentation. Technical Publication No. 69.

ARC 1999: Erosion and Sediment Control Guidelines for Land Disturbing Activities. Technical Publication No. 90. March 1999.

ARC 2000: The Effectiveness of Mulching Earthworked Surfaces. Landfacts S 04. September 2000.

ARC Environment File: No. 14/10 Br 5462 – 64

Environment BOP 2000: Erosion and Sediment Control Guidelines for Forestry Operations, Environment BOP Guideline No. 2000/01.

Goldman S J, Jackson K and Bursztynsky T, 1986: Erosion and Sediment Control Handbook

Herald J R, 1989: Hydrological impacts of urban development in the Albany Basin, Auckland. Ph. D Thesis, University of Auckland.

Humes Industries Limited: Drainage Products Technical Manual

King A N, Greentree D A and Height D R 1998: Construction Site Erosion and Sediment Control. Soil Services, NSW.

New Zealand Forest Service 1980: Civil Engineering Bulletin 4 – Diameter required for small pipes August 1980.

New Zealand Hydrological Society, 1992: Waters of New Zealand.

New Zealand Institution of Engineers, Auckland Branch: A Guideline and procedure for Hydrological Design of Urban Stormwater Systems.

Novotny V, Chesters G 1981: Handbook of Nonpoint Pollution.

Schuler T R 1989: Controlling Urban Stormwater. Washington Metropolitan Water Resources Planning Board.

Soil and Water Conservation Society, Empire State Chapter: New York Guidelines for Erosion and Sediment Control. October 1991.

State of North Carolina, USA, 1988: Erosion and Sediment Control Planning and Design Manual.

Swales A 1989: The effects of urbanisation and consequent sediment generation on the Upper Pakuranga Estuary, Auckland. M Sc Thesis, University of Auckland.

Vant W N, Williamson R B, Hume T M, Dolhin T J, 1993: Effects of Future Urbanisation in the Catchment of Upper Waitemata Harbour. NIWA Ecosystems Consultancy Report No. 220R, prepared for the Environment and Planning Division of the Auckland Regional Council.

Williams P W, 1976: Impact of urbanisation on the Hydrology of Wairau Creek, North Shore, Auckland. Journal of Hydrology Vol. 15, No. 2, 1976.

Chapter 9

Sediment Control – Non Earthworks

9.1 INTRODUCTION

This chapter looks at sediment from sources other than earthworks (eg, from forestry), from various cultivation practices and pasture activities, and finally from channel processes and mass movement erosion. Sediment derived from these sources such as these can make up much of the sediment load of some of the streams and rivers of New Zealand.

Historically, any measures that might have been undertaken to control eroding sites were usually initiated from an on-site perspective. Reasons such as "it will limit production", or "that slip should be plant/grassed" were common in justifying remedial works. The initiating reason could simply be that erosion offended good farmers' sense of stewardship towards their land. Government assistance in the form of subsidies and grants was often available for these works.

Today's perspective, however, can be quite different. Off-site water quality considerations have become much more prominent (particularly with the "effects" base of the Resource Management Act 1991 (the RMA)). Control measures have become much more selective and focused. To some extent this could be because we have more knowledge on various practices and know better their strengths and weaknesses. Some of our traditional practices were just not very effective. More importantly however, government monies has declined on one hand, while on the other there is much more pressure to ensure that adverse sediment related effects do not occur from various activities.

Traditional soil conservation has been almost exclusively undertaken on rural land. Although the terms "sustainability" and "effects" underpin the RMA, it could be argued that historical controls have been on point source discharges whose effects have been easier to control. Activities such as quarries with point source discharges have been controlled for years. Non-point discharges, such as sediment from grazing activities, have always difficult to deal with. Forestry is one of the few rural activities where controls have been imposed for a considerable length of time, but cropping and farming have never been subject to specific controls in the past. Controls are now often for off-site reasons compared to the on-site rationale that was more usual in the past.

9.2 SEDIMENT CONTROL FROM SURFACE EROSION IN PLANTATION FORESTS

In New Zealand approximately 6 percent of the land area is in plantation forestry. Of this area, approximately 90 percent is planted in Pinus radiata. Most of the comments on sedimentation from plantation forestry concentrate on Pinus radiata forests, referred to generally as "pine forests".

Sediment yields are generally accepted to be less than from an equivalent pasture site for most of a forest growing cycle (McLaren 1996). However, there is a short period during logging when sediment levels can be elevated. This is when roads are prepared for logging trucks, landings and skid sites constructed, trees felled and hauled to processing areas. There can be a lot of machinery movement and large areas of land can be bared, particularly as the common method of harvesting in New Zealand uses clearfell techniques. This happens about every 25-35 years

Forest harvesting can produce elevated levels of sediment (Vaughan, 1984), as does land preparation for planting. Harvesting is so closely followed by land preparation that the two phases essentially overlap.

Earthworks (such as roading) have the most impact on sediment yield. Vegetation felling by itself has a much lesser effect on sediment generation and yield than that from earthworks (see Hicks and Harmsworth, 1989). The impact starts when something is done to the vegetation – for example, when the felled vegetation is hauled to some collection point and then transported away. In some places, poor harvesting practices can cause significant sediment generation, particularly on steep slopes and erodible soils. These impacts are usually considered to be less than that from earthworks.

Pine forests that are carefully planted, tended and harvested, can produce forest products for rotation after rotation with minimal generation of sediment compared to other productive uses on the same land. But in many parts of New Zealand, pine forests have been planted on steep erodible land, sometimes as a preferable alternative to pasture. Many of the small forests have had minimal forward planning for future harvesting. These are the forests that have a higher potential for generation of sediment.

Increased levels of sediment arising from forestry operations (other than earthworks) can be a result of:

- loss of protective ground cover during harvest
- exposure of bare ground
- soil disturbance by skidders, tractors or cable haulers
- proximity of operations to watercourses
- concentration of runoff along vehicle tracks and haul paths.

The risk of sediment generation may be increased on:

- steep terrain
- long slopes
- loose soil or weak underlying rock
- timing of operations during wet weather

Production forestry is a commercial enterprise. Harvesting in an environmentally sensitive manner can be more expensive. Where forests are planted on steep land, there are limited options for harvesting. Often, there will be some unavoidable adverse environmental effects. The following sections deal with ways of minimising sediment generation during forestry operations.

9.2.1 Principles of erosion and sediment control

Plan ahead to minimise erosion

The most important aspect of forward planning is to tailor the planting boundaries to match an appropriate proposed method of harvesting. The proposed harvesting system will dictate how much roading will be required, and the general layout of landings, haul tracks etc, is critical in minimising sediment generation. Forward planning of forestry operations is the most effective method of reducing problems. This needs to be followed up with good operational practice, and regular monitoring and maintenance once the works are completed.

Minimise bare ground as far as practicable

Ground cover helps to prevent erosion of the soil surface. Some land preparation for planting (such as burning or desiccation) can completely destroy all of the surface vegetation and expose the bare ground to the erosive forces of the weather. Sometimes, the operations are undertaken to prolong the exposure of bare ground so that the new seedlings have reduced competition from weeds. When afforesting steep erodible country, consideration should be given to alternative methods of land preparation that minimise the exposure of bare ground. This could involve oversowing with grasses and legumes.

Minimise soil disturbance as far as practicable

Many forestry operations involve soil disturbance even though they are not classed as earthworks. Such operations include stumping, hauling logs, ripping, V-blading, and root raking. These operations can often be undertaken with minimal generation of sediment, provided they are properly planned and supervised. Where there is a higher risk of adverse effects, such as on steeper country or near streams, these operations should be avoided if possible.

Protect riparian areas, and don't deposit soil or debris into watercourses

Riparian areas are critical because any discharge may have adverse effects on the downstream water resources. Forestry operations carried out close to streams, or over streams, should be undertaken carefully so that any deposition of soil or debris into the watercourse is avoided. Where there is an increased risk of sediment entering a stream, it may be prudent to designate a protected area next to the watercourse. The accumulation of waste logging debris ("bird's nests") at landings can also lead to problems if the landings are sited above steep faces, and the bird's nest collapses.

Avoid the concentration of stormwater runoff

Any operation that concentrates stormwater runoff increases the risk of erosion. Some of these are cultivation/ripping operations that run downhill rather than on the contour, hauling logs downhill to a common point, or locating skid sites in a low point or gully etc. Often, the effects of concentrating stormwater runoff are not evident until after heavy rainfall. These activities can be managed to reduce volume and velocity of stormwater runoff.

9.2.2 Forest planning

Forward planning includes planning to establish a forest from scratch, as well as before each stage of the forest rotation. The New Zealand Forest Code of Practice (LIRO 1993) is a valuable tool for assisting forest owners/contractors in identifying any adverse effects of a particular operation, and then in selecting the appropriate techniques to reduce potential impacts.

Inadequate forward planning often results in higher cost in the long term. Professional forestry consultants have the expertise to carry out forestry planning as well as supervise forestry operations. While it is essential to have a good contractor carrying out the forestry operations, a good operator cannot make up for poor forest planning. For example, a good ground based logging operator will have difficulty logging a steep site that should be logged by cable hauler.

9.3 SEDIMENT CONTROL FROM SURFACE EROSION IN CROPLANDS

Cultivation of soil is necessary to create a better seedbed, and to minimise weed competition on growing crops. It usually exposes the soil to the erosive effects of rain splash and concentrated runoff. The potential for erosion and sediment generation from a cultivated site can be far greater than from the same site under a permanent vegetative cover. Loss of soil from cultivated land can result in elevated levels of sediment in receiving waters and also impact on the sustainability of the soil resource itself.

Three different types of cultivation on three different kinds of soil are discussed here:

- market gardening on granular loams
- cropping on loess
- orchard cultivation on free-draining alluvial soils.

9.3.1 Market gardening and grain cropping on granular loams

Granular soils derived from volcanic ash are some of the most intensively used in New Zealand. Approximately 8,000 ha are used for large-scale vegetable production in South Auckland (Franklin District) (Basher, 1997), and there are smaller pockets in Northland, North Auckland, and coastal Otago. Anecdotal evidence suggests that severe erosion can occur on these soils during intense rainfalls and that this can not only adversely affect the sustainability of the soil but also lower water quality.

The Auckland Regional Council undertook a catchment study into sediment yield from vegetable growing in the early 1990s. Sediment yield from the 1.8 km^2 trial catchment site was 49 t/km^2/yr over three years (Hicks, 1995). Bedload was negligible at the sampling site. Soil loss from earlier plot studies undertaken by Cathcart et al (unpubl) in the early 1970s was evaluated (Basher et al, 1997). Bare soil plots averaged 5680 t/km^2/yr over a 2.5 year measurement period. The largest storms resulted in most of the soil loss mentioned in the two studies.

The very large difference in yields suggested that large quantities of soil were being mobilised within paddocks by storms but little was being transported as suspended sediment load by streams. This was considered to be a function of local drainage characteristics (whether fields discharge directly into watercourses or not) together with the strongly aggregated nature of the soil. Analysis of the soil indicated that the soil is resistant to dispersion into primary sand, silt and clay particles (silt and clay sized particles usually constitute the suspended sediment load). It was inferred that the soil which is moved in storms is transported as aggregated material and deposited within paddocks and drains when runoff velocities diminish.

When a regional plan specifically directed towards controlling sediment from the Auckland region's land disturbing activities was prepared and notified in 1993, the council did not impose any regulatory controls on outdoor vegetable production, as the catchment soil loss yield of 49 t/km^2/yr was similar to that from pasture from other soil types in the region (Hicks, 1994). However, opinion continued to hold that significant soil loss was occurring under this type of land use and that this was affecting both soil sustainability and water quality.

A major storm in May 1996 resulted in flooding and sediment deposition in parts of Pukekohe township and focused public attention on the practices of outdoor vegetable growers. Discussions were then held between the Pukekohe Vegetable Growers Association, regional councils and Agriculture New Zealand, and as a result the Franklin Sustainability Project came into being. This was a three-year project with a emphasis on erosion management, although wider considerations, such as irrigation, pest control and nitrate leaching, were also included. Much of the following information has been obtained from this project.

Sheet erosion of topsoil following a storm event, Pukekohe, 1999. *Photo: D. Hicks.*

9.3.2 Principles of control on granular loams

Granular loams are strongly aggregated, and the aggregates are not easily broken. They do not break down to fine clay, silt and sand-sized particles but to smaller aggregates. The soil aggregates are still relatively large compared to clay and silt, and can quickly settle out once flow velocities diminish. Under normal conditions, infiltration and percolation through these soils is very high. Measures to ensure that aggregate stability continues are important.

This characteristic markedly influences the type of erosion and sediment control measures that can be used. Soil particles may be relatively easily mobilised, but they are also equally easily retained once flow velocities diminish. Effective erosion and sediment control does not then merely consist of providing settling areas for suspended sediment to settle out. It comes down to minimising off site runoff and reducing on site runoff.

For each paddock, the following points need to be checked:

- Where is the water coming from?
- Where is the water going?

- How can the paddock be set up to minimise erosion and soil loss?

Six basic measures are promoted to minimise erosion and sediment loss:

- channels and interception drains to intercept above-site runoff
- benched headlands
- permanent drainage systems within paddocks
- contour drains
- raised accessways
- silt traps.

The first five are effectively runoff control measures while the last is the only sediment control measure. In addition to these measures, a number of management practices have also been suggested. These include ripping wheel tracks, growing cover crops, use of hedges and cultivation techniques.

A paddock plan will help ensure that erosion and sediment control measures are appropriate for the site and mesh in with growing practices. It should be integrated with other paddock plans to form a co-ordinated property plan, vital for runoff control. Each property plan should be integrated with adjacent property plans, where they exist.

Each measure or practice (except for permanent drainage within paddocks, which is self-explanatory) is briefly discussed below in section 3.1.3.

9.3.3 Grain cropping

Cultivation of arable loess soils in the eastern South Island and southern North Island can result in the loss of soil by water runoff and by wind erosion. This soil loss is a consequence of most of loess soils high susceptibility to structural degradation. Cultivation can readily break the soil down into individual silt grains, susceptible to erosion by both water and wind.

In Canterbury and Otago, arable soils are considered to be at more serious risk of wind erosion than water erosion (Ross et al 2000), due to strong, drying foehn winds (nor'westers). However, a report by Landcare Research (Ross et al 2000) prepared for the Canterbury Regional Council assesses the risk of erosion on arable soils; it also addresses soil loss by surface runoff and has been used extensively in compiling the following information.

Sheet and rill erosion occur when water runs off cultivated topsoil on sloping ground. Low permeability subsoil in loess limits downward water movement, and sideways flow through topsoil contributes to runoff generation at low points in fields. High erosion rates on cultivated downlands have been recorded since the 1940s and found to be closely associated with recent cultivation (Ross et al 2000).

Rill erosion is considered to most significant with up to fifty times more soil being lost through rilling than from other forms of erosion (ORC, 1995). Rill erosion induced by moderate intensity rainstorms on finely cultivated soils can deposit 20–93 t/ha of sediment at the foot of cultivated slopes (Hunter et al, 1990). Rill dimensions indicated local losses on eroded slopes of up to 410 t/ha. Hunter and colleagues also reported an earlier study of soil loss from rill erosion in the order of 50 to 101 t/ha for four sites. During a three month period in 1986, 240 ha over 24 properties lost an average of 50 tonnes of topsoil/hectare (ORC, 1995). The depth of topsoil after 60 years of intensive cultivation was found to be 100 mm shallower compared to equivalent uncultivated slopes (Hunter et al, 1990).

While mean topsoil depth did not differ significantly between areas used for pasture and areas used for cropping, topsoil depth under cropping was far more variable and there was a higher frequency of shallow topsoils (Basher et al, 1995). Basher's study found considerable redistribution of soil by erosion and deposition but little export of soil. Ross et al 2000 support this and note that erosion on downlands results in considerable redistribution of soil but little net loss. The likely reason is that there are few permanently flowing watercourses close by, to transport eroded sediment off-site.

Rilling is often worse after a drought period when soils are dry. Storms can therefore mobilise and transport large quantities of soil from land under cultivation. Storms of sufficient intensity to cause significant erosion can be expected from the larger rainfall events, such as those with a greater than a 10year return period. Most of this soil is deposited at the foot of slopes and can cause localised problems to occur such as blocked culverts etc.

9.3.4 Principles of control

Control of water-initiated erosion on cultivated soil includes measures that increase soil resistance to erosion and those that reduce runoff volume and velocity. The condition of the soil, the rate of infiltration into the soil, the presence of subsurface pans (both natural and from cultivation), poor vegetative cover, and long steep slopes, are all factors that influence erosion on cropping areas. Retaining or improving the structural stability of the soil is a key element of control.

9.3.5 Orchards and vineyards

In the 1960s, apple and pear orchards within the Nelson Province were suffering from severe erosion problems. At that time, the Nelson Catchment Board became heavily involved with promoting erosion control works and developing practices to prevent future problems. The Board was assisted with funding from the Soil Conservation and Rivers Control Council to develop sustainable land management practices using six demonstration orchards. By 1977, effective practices had been developed and a review of the conservation works was published (Leighs, 1980). The information below is from that review.

The problems stemmed from a number of factors:

- The Moutere Soils are relatively thin and non-cohesive with little or no organic matter. The subsoil is hard clay over gravels.
- The orchards were often planted on sloping ground in a square pattern using straight rows that is best suited to flat land.
- Rainfall in the Moutere Hills area averages 950 –1050 mm/year, with dry summers and mild winters. Storms of 70 mm in 24 hours, having a return frequency of 2 years, can occur at any season.
- In an attempt to conserve moisture and carry out weed control, the ground was cultivated between the rows of trees.

Cultivation resulted in the ground being kept tilled from mid spring to mid autumn, with many owners priding themselves on the fine tilth achieved. Over time, excessive working of the soils resulted in a cultivation pan on the already hard subsoil and weakened the poorly structured topsoil. As a result, even light rains produced widespread sheet and rill erosion. Heavy rainstorms moved large quantities of soil from the upper slopes to lower ground and valley floors. It was common for the roots of fruit trees on spurs and upper slopes to be left standing on pedestals of earth, because the soil between the trees had been washed downslope to deposit on the flats and create swampy conditions burying trees and fences.

9.3.6 Principles of control

The key principles for control of the problems included:

- improving infiltration of rain and reducing overland flow runoff
- controlling subsurface and surface water, and conveying the water safely to lower flats and natural waterways
- stabilising exposed ground with grass.

The practices that were developed to address the above principles were a combination of runoff control, drainage control and erosion control.

9.3.7 Practices

The pattern of works included the following techniques:

- **Ripping of subsoil** to promote infiltration of rain and reduce runoff;
- Installation of **lateral grassed waterways** with subsurface tile drainage to collect underground and surface water;
- Installation of **main grassed waterways** with subsurface tile drainage to receive water from laterals and some direct subsoil ripping;

- Use of main **outlet grassed waterways** with subsurface tile drainage to receive water from several main tiled grassed waterways and deliver it to;
- Main **outlets or disposal systems**, usually natural streams, but could also be detention or retention ponds;

Grassing of all exposed ground and drainage systems.

9.4 SEDIMENT CONTROL FROM SURFACE EROSION IN PASTURE

Surface erosion of topsoil causes reductions of 40 percent or more in pasture growth. The severity of the impact depends on the frequency of erosion and how much land is affected. Surface erosion occurs on an annual, or more frequent, basis. In low rainfall areas where vegetation is often sparse, up to half the land can lose some of its topsoil each time, but in high rainfall districts which are usually well vegetated, topsoil loss is restricted to a tenth or less of the surface.

Farming practices that may inadvertently cause surface erosion are those that deplete ground cover. Windblow, sheetwash and rilling of topsoil can occur almost anywhere where groundcover is depleted, even on flat land. Light soils with a high silt or sand content are more susceptible than loams, heavy clays or peats.

Other practices that will lead to sediment generation include those that weaken soil strength, by physical destruction of the soil aggregates, or breakdown of the soil structure.

Very often, erosion can be aggravated or induced by stock management. In a literature review on the adverse effects of grazing on soils, Hunter-Smith (2000) noted that overgrazing can adversely affect soil properties, particularly infiltration rate, macroporosity and bulk density. The extent of the damage is influenced by a number of factors:

- slope;
- soil type, texture, and moisture content
- pasture cover and species
- stocking rate, duration, and animal species.

A number of studies conclude that winter and spring appear to produce the highest amounts of runoff and sediment transport. However, other studies also show that when the soil is dry (generally in summer), surface runoff and soil erosion increase during high intensity rainfall. Maintaining a dense grass sward and avoiding overgrazing, protect the soil surface from very intense rainstorms, particularly those that occur during summer. Retirement from grazing significantly decreases surface runoff and increases infiltration rates.

9.4.1 Principles of control

The following principles cover the control of sedimentation from pastoral land use:

- Avoid land management practices that deplete ground cover or expose bare ground / Adopt practices that encourage a dense or close ground cover.
- Avoid land management practices that weaken soil strength or break down soil structure.
- Encourage land management practices that increase infiltration capacity of the soils.
- Manage streamside areas carefully, and protect them whenever possible.

9.4.2 Practices

Any farm or pasture management techniques that minimise bare ground and help form a dense pasture sward will reduce the risk of surface erosion, limit adverse effects on soil structure, and help prevent a decrease in the rate of infiltration.

Application

While it is accepted that overgrazing of pastures cannot be completely avoided, all pastoral farmers will attempt to maintain an adequate ground cover, and minimise the risk of overgrazing as far as practicable. There are a range of different farm management practices that can be undertaken by landowners that minimise the amount of bare ground, and help encourage a dense pasture sward. These include the following:

- Have the farm well fenced so that stock grazing can be closely controlled.
- Have a reticulated water supply that is well planned to provide adequate water to stock, particularly during peak demand periods such as mid summer. Ensure that there are sufficient troughs available, so that grazing animals do not have to travel far to the nearest watering point.
- Establish a pasture improvement programme to provide a dense pasture sward, particularly on pastoral hill country.
- Provide an adequate fertiliser programme to maintain a good pasture sward.
- Avoid mob stocking or heavy set stocking during periods of drought, cold conditions or wet weather.
- Carefully site of fences, gates, troughs and farm tracks to avoid low points or overland flow paths where stormwater runoff may be concentrated.
- Keep stocking rates to a sustainable level for the classes of land that are being grazed, particularly during wet periods.
- Use grazing pressure as a pasture management tool without overgrazing.
- Match the type and class of stock to the land capability of the property.

In general, good pastoral farming practices will also help retain a dense pasture sward, and will minimise bare ground and reduce the risk of sedimentation.

9.5 SURFACE EROSION CONTROL FOR OTHER SOURCES OF SEDIMENT

Sediment can arise from sources other than the commonly recognised ones such as earthworks, forestry, and cultivation. Sediment from mass movement erosion and stream channel processes are two.

In some areas around New Zealand, mass movement erosion is occurring on a scale that results in large quantities of sediment being regularly discharged to waterways. This sediment can then be reworked on an ongoing basis and the receiving environments may have little chance of recovery. These situations are therefore very different one-off type sources of sediment such as earthworks.

Much of the emphasis on mass movement control has been related to the on-site effects. Less prominent, but also relevant, is the effect of the ongoing injection of sediment into the receiving environments. Not all eroded material enters watercourses, but what does can have devastating effects. If sediment continues to enter watercourses, or ongoing processes continue to rework sediment deposited between regular storm induced “top-ups”, then the receiving environment may not be able to recover to its former level. The effects of mass movement erosion in these situations then not only relate to the sustainability of the on-site landuse, but also to the off-site sustainability of the receiving environments. A permanent degradation of the receiving environment can occur.

An example of this is in the siltstone country north of Napier. There are a considerable number of studies that have assessed the impact of storms and subsequent erosion in this area of New Zealand. One of these assessed the soil mobilised by a five year return period storm to be nearly 9000 $t/km^2/yr$ (Eyles, 1971). Storms of about a 5-year return period interval appear to be about the threshold for significant mass movement erosion. Page (1992) assessed that Cyclone Bola, which occurred in March 1988 and was the largest rainfall event to be recorded in the area, generated an average of 420 m^3/ha (42,000 $m3/km^2$) within the same catchment. Of this sediment generated, 51 percent was deposited on lakebeds and 6 percent was discharged from the lake outlet. The study considered that most of the sediment generated from this storm (89 percent) was from landslide erosion, and that sediment in secondary storage provided only a small contribution. The sediment deposited by larger storms was reworked within the streams and rivers (Page et al, 1994), during rainfalls less than the 5-year threshold.

Not all sediment generated by mass movement erosion enters a watercourse. The proximity to watercourses, slope, surface roughness, retaining vegetation, and the characteristics of the drainage pattern will all influence how much is retained on hill slopes and how much discharges to watercourses. For instance, about 50 percent of sediment generated by landslides in the Tutira area was considered to enter watercourses (Page 1992). The same report also assessed a number of other studies around New Zealand and concluded that, even with different terrain and triggering conditions, about 50-55 percent of all sediment generated from mass movement sources entered streams. This will obviously vary however, for the reasons mentioned above.

9.5.1 Sediment from channel erosion

The Alluvial Process

A watercourse develops a bed shape in response to the volume of water that the watercourse must carry, the size and volume of sediment carried, the hardness of underlying rock, the slope of the channel and the "age" or geomorphological stage of development of the catchment and its floodplain (Cathcart, 2000). A regime, which generally includes a meandering channel, some channel erosion and deposition of sediment, is just one variant. If any of the catchment characteristics are changed, the watercourse will accordingly adjust its regime to compensate. The channel may alter –for example, to a straight or braided form – as it accommodates the changes, with bank erosion and deposition also altering.

Channel erosion can contribute to undercutting and instability of adjacent land. This in turn can affect sediment generation and supply to the channel. Observation by river engineers confirms that channel instability can sometimes be related to a sudden introduction of detritus to the stream. This sets up a chain reaction, by initiating bank erosion, leading to more bedload, leading to more bank erosion. Once started, this process can be difficult to control, as it requires control over the input of material, as well as control over bedload movement and bank erosion.

Bankfull or near-bankfull is generally considered the most effective discharge in transporting sediment and reshaping the channel bed and banks. (Waters of New Zealand, 1992). In many streams this is about the mean annual flood event (this is the average of the series of annual maximum flows, which statistically has an average recurrence interval of 2.33 years – Waters of New Zealand 1992). Once eroded, deposition of clay and other fine grained materials is unlikely in streams because of a low settling velocity. Concentrated flows in channels have greater depth and velocity and sediment carrying capacity than overland flow. As a consequence most fine textured sediment derived from stream bank erosion will be reworked downstream over time.

Channel Erosion as a Source of Sediment

Stream channel erosion is a common form of erosion and can be a major source of sediment. From an "effects" perspective, sediment from channel erosion is immediately available for transport by channel flow. It is therefore different to catchment derived material may only travel short distances and can be re-deposited and retained in the catchment. A good illustration of this is the relatively small quantities of sediment that might leave a vegetable-growing area on granular soils in South Auckland compared to the very much larger quantity that can be generated within the catchment.

The contribution of sediment from channel erosion varies tremendously. Gianessi (1986) reports that a study of sediment sources in five northern Mississippi watersheds found channel erosion accounted for 15 to 28 percent of the total sediment delivered while being only 1 percent of the catchment area. Channel erosion can therefore be an important source of sediment and can have an influence on water quality quite out of proportion to the areal extent of watercourses.

The relative influence of channel erosion as a source of sediment varies similarly in New Zealand. In one study (Page et al 1992) channel erosion was considered to contribute only 2 percent of the total sediment yield from one catchment (Lake Tutira, north of Napier) during one storm although there was significant channel modification. Little sediment was found to accumulate between the larger storm events and this suggested that sediment from channel erosion in this catchment was significantly less that from mass movement erosion.

9.6 THE EFFECT OF URBAN DEVELOPMENT ON CHANNEL SEDIMENT

The change of landuse from rural to urban increases the imperviousness of a catchment surface and changes its hydrological characteristics. Increases in flood-flow frequency and size result from the changed runoff patterns. Stream channels will widen as they re-adjust to the new flow regime. Herald (1989) stated that there was a three fold increase in the width of streams draining an area that was 85 percent urbanised. Schueler (1989) supports this by claiming that most streams widen two to four times their original size if post-development runoff is not effectively controlled. These changes can start to occur once 10-20 percent of a catchment has been developed. This stream widening is primarily accomplished by lateral cutting of the stream bank resulting in undercutting of the zone adjacent to the channel, treefall and slumping. This channel widening occurs in direct response to changed flow regimes resulting from urbanisation. Flood detention measures in some urban situations throttle flood flows down to the bank full situation and which, although controlling flood peaks exacerbates channel erosion because the bank full situation (the main erosion causing flow) is continued for a longer period.

One study looking at the effects of proposed urban development predicted an average of 9,800 tonnes/year of sediment would result from widening of channels in one catchment in Auckland as the channels adjusted to altered flow regimes over a 20-year development period (from Vant et al, 1993). This sediment source was solely due to streams widening and to accommodate altered hydrology resulting from changed catchment characteristics. This was compared to 18,000 tonnes/year estimated from the earthworks phase of development (not all of which would reach watercourses), and 2,800 tonnes/year from the remaining catchment sources such as pastoral and developed urban land uses.

Channel erosion resulting from urbanisation of catchments can therefore be a major contributor of sediment to waterbodies. Modification of channels will normally last for years as stream channels adapt. The generated sediment is generally immediately available to stream flows and can have effects on receiving environments out of proportion to its relative area.

9.7 REFERENCES

RC undated: Sediment yield from channel works on the Papakura Stream. (unpublished)

Basher LR, Mathews KM and Zhi L, 1995: Surface erosion assessment in the South Canterbury downlands, New Zealand using 137Cs distribution. Australian Journal Soil Research, 33, p 787-803.

Basher LR, Hicks DM, Handyside B and Ross CW, 1997: Erosion and sediment transport from the market gardening lands at Pukekohe, Auckland, New Zealand. Journal of Hydrology (NZ) 36(1) p 73-75.

Cathcart R, 2000: Controlling stream channel erosion in Northland. Soil Conservation Fact Sheet 1, Northland Regional Council 2000.

Collier K, M Simons, P Simpson, J Waugh 1992: Riparian zones and sustainable land management – A DOC perspective. Broadsheet Spring 1992 New Zealand Association of Resource Management (NZARM).

Environment BOP 1998: Sustainable Options Fact Sheets SC02/98:Riparian retirement; SC07/98:Plant selection for retirement areas; SC08/98:Management of retirement areas; SC09/98:Runoff management on pastures.

Environment BOP, 2000: Erosion and Sediment Control Guidelines for Forestry Operations, Guideline Number 2000/01, July 2000.

Evans B 1983: Revegetation Manual: A Guide to revegetation using New Zealand native plants. Queen Elizabeth II National Trust.

Eyles RJ, 1971: Mass Movement in Tongoio Conservation Reserve, Northern Hawkes Bay. Earth Science Journal, vol 55, No. 2.

Franklin Sustainability Project, 2000: Doing it right. Guide to Sustainable Management.

Gianessi L P, Peskin H M, Crosson P, Puffer C, 1986: Non-point source pollution: are cropland controls the answer? Journal of Soil and Water Conservation, July-August 1986.

Hicks DM 1994: Storm sediment yields from basins with various landuses in Auckland Area. ARC Environment Division, Technical Publication No 51, Auckland Regional Council, Auckland 34 p.

Hicks DM 1995: Sediment yields from a market gardening basin near Pukekohe. ARC Environment Division, Technical Publication No 61, Auckland Regional Council, Auckland 24 p.

Hicks DL 1995: Control of soil erosion on farmland. MAF Policy Technical Publication 95/4, Ministry of Agriculture and Fisheries. 45 p.

Herald J R, 1989: Hydrological impacts of urban development in the Albany Basin, Auckland. Ph. D Thesis, University of Auckland.

Hunter GG, Lynn IH 1990: Storm induced soil losses from South Canterbury and North Otago downlands. D.S.I.R. Land resources Technical record 22. 7 p.

Hunter-Smith J 2000: Draft Report – The Adverse Effects of Grazing on Soils – A Review of Literature. Environment BOP, Environmental Report 2000/14, July 2000.

Leighs MJ 1980: A Review of Water and Soil Conservation Works in Nelson Orchards. Nelson Catchment Board and Regional Water Board.

Maclaren JP 1996: Environmental Effects of Planted Forests. FRI Bulletin No. 198, New Zealand Forest Research Institute.

Murphy G 1992: Riparian Zone Management – A Review of the Literature. Prepared for the Bay of Plenty Regional Council.

New Zealand Forest Research Institute 1992: Proceedings of Riparian Zone Workshop – Implications for Exotic Forestry, compiled by JA Fenton.

NIWA 1994: Managing riparian zones along agricultural watercourses. I. Background concepts, and II. Interim guidelines. National Institute of Water and Atmospheric Research Limited, Hamilton.

Otago Regional Council 1995: North Otago Sustainable Land Management Guidelines.

Page MJ, Trustram, NA and Dymond JR 1992: Sediment budget to assess the geomorphic effect of a cyclonic storm, New Zealand. Geomorphology 9 (1994) p 169-188.

Page MJ, Trustram, NA and DeRose RC 1994: A high resolution record of storm induced erosion from lake sediments, New Zealand. Journal of Paleolimnology 11, p 333-348.

Quinn JM 1994: Framing the Issue: Why Are Riparian Areas so Important? Paper presented to NZARM Annual Conference 1994.

Ross CW, Basher LR & Baker CJ 2000: Management of erosion risk on arable soils: a review. Landcare Research Report LC9900/090 prepared for Environment Canterbury regional Council. 119 p.

Schuler T R 1989: Controlling Urban Stormwater. Washington Metropolitan Water Resources Planning Board.

Shepherd TG, Ross CW, Basher LR and Sagger S 2000: Soil management guidelines for sustainable cropping. Landcare Research.

Spiers JJK 1985: Loggers Language A New Zealand Terminology. New Zealand Logging Industry Research Association Inc. February 1985.

Spiers JJK 1986: Logging Operations Guidelines. New Zealand Logging Industry Research Association Inc. and National Water and Soil Conservation Authority.

Stringer OA 1978: The river scene: The downstream viewpoint. Paper presented to conference on "Erosion Assessment and Control in New Zealand", New Zealand Association of Soil Conservators 25th Anniversary Conference, University of Ilam, Christchurch August 1978.

United States Department of Agriculture 1954: A manual on conservation of soil and water – Handbook for professional agricultural workers. US Department of Agriculture, Soil Conservation Service Agriculture Handbook No.61.

Vant W N, Williamson R B, Hume T M, Dolhin T J, 1993: Effects of Future Urbanisation in the Catchment of Upper Waitemata Harbour. NIWA Ecosystems Consultancy Report No. 220R, prepared for the Environment and Planning Division of the Auckland Regional Council.

Vaughan L 1984: Logging and the Environment. NZ Logging Research Association

Vaughan L 1990; Visser R and M Smith 1993 (eds.): New Zealand Forest Code of Practice, 2nd Edition. New Zealand Logging Industry Research Organisation.

Wendell G, L Schipper, P Beets, M McConchie 1992: Riparian Buffers in New Zealand Forestry. New Zealand Forest Research Institute.

Wilkie DR: The curving furrow – Soil conservation by contour methods. Bulletin No. 10. Soil Conservation and Rivers Control Council, New Zealand.

Williamson RB, CM Smith, AB Cooper 1996: Watershed Riparian Management and its Benefits to a Eutrophic Lake. Journal of Water Resources Planning and Management January February 1996.

Part - B

Practices

Surface Erosion Control

Chapter

1. Cropland Management

2. Pasture Management

3. Fencing Management

Chapter - 1

Cropland Management for Surface Erosion Control

1.1 INTRODUCTION

These management practices encompass conventional tillage, minimum tillage, no tillage, stubble mulching and contour cultivation. Runoff management practices for arable land are described separately (see 10).

It is important to understand the terminology used with respect to cultivation, minimum tillage and direct drilling in NZ.

- Conventional Tillage – Is the practice of establishing a crop or pasture in the field which has been tilled or cultivated (Includes the full cultivation process to seed bed stage).
- No Tillage – To establish a crop or pasture where no cultivation is carried out at all. This term is synonymous with Direct Drilling.
- Minimum Tillage – is where the soil surface may have only one or two passes over it by using say harrows or light discing then sowing the seeds (rather than full cultivation process).

It is important to note that most cultivation equipment will mechanically move soil downhill leaving upper slopes and knobs bare of soil. Erosion greatly speeds up this movement, which can be clearly seen on paddocks that have been ploughed one way for many years.

1.2 CONVENTIONAL TILLAGE

1.2.1 Guidelines

Many farmers prefer conventional tillage with a plough, prior to seedbed preparation with harrows or discs. Downslope soil loss can be reduced by the following practices:

- Cultivating on the contour where possible
- Maintaining slow tractor speed reduces the distance soil is moving downhill

- Travel slower across steeper faces and if going downhill on gentle slopes using rotary hoe, maxitill or grubber (as this equipment can move a wave of soil downwards if travelling at speed)
- Cultivating at the correct time i.e. when soil is moist and friable to minimise the number of passes required
- Use a reversible plough to throw the furrow uphill
- Use other equipment that will have the least impact i.e. paraplow or Maxitill in preference to rotary hoe (i.e. over-use of a rotary hoe will damage the soil structure)
- Where possible use single pass cultivation with multiple implements attached, or a one pass cultivator; may need higher HP tractor or alternatively use a contractor with the specialised equipment
- Judicious use of suitable herbicides may reduce the number of passes required
- Where possible avoid cultivating slopes over 13 degrees especially if the slopes are longer than 30 metres
- Minimise the bare ground by timing the tillage operation (consider rotation type and cover crops)
- Avoid leaving a fine seedbed (surface roughness is vital with a range of aggregate sizes which will promote better infiltration thus reducing the surface runoff and erosion).
- Take care to minimise compaction from machinery by not cultivating in winter months, or when the soil is wet in spring and autumn
- Leave dry, ephemeral watercourses uncultivated i.e. in permanent pasture – Refer to Grassed Waterways.
- Plant steeper vulnerable slopes in trees and fodder banks.

Steeper area in poplars – no fencing. Cultivation on more gentle sloping land.
Photo: M Harris.

Finally if cultivation is deemed to be a necessary practice on the farm then it is vital that ongoing monitoring of the soil aggregate stability is carried out e.g. by visual soil assessments, Shephard (2000).

1.3 MINIMUM TILLAGE

This tillage practice has been used in New Zealand for many decades, more so before direct drilling technology was mastered. There has been a significant move into minimum tillage in the areas of New Zealand commonly affected by droughts i.e. the eastern side of the South Island (especially after the drought of the eighties and late nineties), also Wairarapa and Hawkes Bay in the North Island.

Minimum tillage may involve using simple cultivation equipment and passing over a paddock only once or twice e.g. with harrows then broadcasting or drilling the seeds. The more common approach today is to use an approved herbicide to reduce the vegetative top cover thus reducing the energy (i.e. fuel used) and machinery passes to prepare a seedbed.

There are advantages over direct drilling, such as minor levelling of a paddock and generally lower cash costs and machinery lasting longer.

1.3.1 Guidelines for minimum tillage

Aim to only carry out sufficient cultivation to prepare the paddock for drilling (use suitable equipment as a maxitill or one way pass cultivator)

- Leaving a rough cloddy seedbed and cultivate to a shallow depth.
- Use approved herbicide to kill existing vegetation (may need two sprays depending on weeds present) this also helps the ground break up, conserves soil structure and moisture. Note that if the soil structure is fine with poor aggregates then it is preferable to direct drill
- On slopes of 13-20 degrees, minimum till only when coming out of long term pasture. If sowing Brassicae crops then include suitable grass species in the mix for good ground cover
- Take care not to get wheel compaction on clay soils

1.4 NO-TILLAGE CULTIVATION

Modern technology developed in New Zealand has provided sophisticated coulter designs such as the "cross slot drill" which have eliminated many of the earlier problems and variable results with traditional drills when sowing pasture and crops.

Some of the benefits of direct drills compared to cultivation can be summarised as follows:

- Minimal surface erosion of topsoil

- Soil improvements as the soil structure is preserved and improved whilst leaving pasture and crop residues including stubble on the surface
- Enhances earthworm numbers and other soil fauna
- Fewer new weeds introduced
- Time savings in the number of passes
- Fuel conservation up to 80% less fuel used and lower machinery costs
- Reduced nitrogen mineralisation and carbon oxidation to the atmosphere

Such cultivation systems are already well established in many countries, such as the USA and Canada, and parts of Europe. The benefits and several of the risks have been described in books such as "No-Tillage – Science and Practice" by C J Baker, K E Saxton and W R Ritchie (1996), and "Successful No-Tillage – Practice and Management" by W R Ritchie, C J Baker and M H Hamilton-Manns (1998. In addition, the North Otago Sustainable Land Management Group has a very useful overview of "Direct Drilling and Pesticide Use" at www.noslam.co.nz/info_sheets/directdrill ing.shtml See also "North Otago Sustainable Land Management Guidelines, produced by Otago Regional Council, (1995)".

Choice of drill

In general, the more recent the design of direct drills, the better the results. New triple disk designs with small press wheels give better results than older designs. Triple disc drills perform best when moisture conditions are good, or in sandy or stony soils.

1.4.1 Tips for successful direct drilling

The following steps are recommended for successful direct drilling:

- Test soil well in advance. If lime is required, put it on 6-12 months prior to drilling. Identify run out paddocks in May/June.
- Spray old pasture while soil moisture levels are reasonable and when the pasture and weeds are actively growing, usually between September and November. An excellent plant kill is essential – seek good advice on correct chemicals.
- Where moisture retention is a problem, use the double spray – fallow technique to improve moisture availability. Trials have shown that up to 12 times more moisture is retained to 20 cm depth compared to unsprayed sites. Old pasture killed with herbicide [or other measures], followed by fallow for at least 6 weeks, does not transpire and thus retains soil moisture. Plant roots also break down.
- Inspect newly drilled paddock every 2-3 days while seed is germinating and seedlings are becoming established; take action if insect, slug or weed problems occur.

- Correct timing. If you can wriggle a pocketknife into the soil up to the end of the blade, soil conditions should be suitable for direct drilling. If not, it is either too dry, or the roots have not broken down sufficiently.
- Do not direct drill when conditions are too wet. Smearing of soils can result if conditions are not suitable; this will drastically affect germination.
- Best sowing depth. Sow no deeper than 13 mm for most seeds. If sown deeper, the establishment rate for most pasture plants will drop sharply.
- Harrowing, if the drill used did not have its own press wheels, harrowing helps to produce loose soil to cover the seed in the slots, but be very careful to avoid soil compaction at the seed slots. That would also reduce the establishment rate of the pasture plants.

1.4.2 Results of direct drilling

Ross (2000) found that soil organic matter levels in the top 20 cm under long-term (21 years) no-tillage in the Manawatu were maintained at 85-94% of those under permanent pasture. This compared with a decline of 40% in soil carbon after 17 years of cropping by tillage. Total weight of earthworms under long-term no-tillage cropping was 77% of that under pasture, which compared with a 99.8% decline under 29 years of tilled cropping. The aggregate stability of the no-tilled soil was 85-92% of that under pasture, and five times better than that after 23 years of continuous cereal cropping by tillage. Over the years no-tillage soil became progressively easier to drill (requiring decreasing levels of tractor power) which has been explained by the soil's increasing crumb structure and aggregate stability. Such an observation has since also been confirmed in Canterbury (D, M and A Redmond; R and M Scott; D Ward, pers. comms (1999), in Baker (2000) b).

1.4.3 Pest monitoring and management

During conventional cultivation, plant pests may be mechanically damaged or more easily found by various birds, such as gulls and starlings. With a direct drilling programme, this does not occur and the risk of pests damaging crops increases. Control is possible, but it does increase the use of chemicals. But there are a number of practices that can be used to minimise the use of chemicals in the pest control programme.

Recommendations for pest monitoring and management are:

Grass grub

Dig 10-15 inspection holes across a paddock. If there are more than two grubs per spade square, insecticide should be used.

Springtails

Springtails damage brassica crops soon after sowing, often before seedlings emerge. To establish if springtails are likely to be present at damaging levels, place a white handkerchief on the soil in the afternoon, seven days after sowing. If more than five springtails have jumped onto the handkerchief after 2-3 minutes at ten sampling points across the paddock, then damaging levels are probably present. But spray only if springtail levels are high enough.

Argentine Stem Weevil

Argentine Stem Weevil [ASW] affects all improved pasture and damages by larval feeding and stem mining. ASW resistant ryegrass "high-endophyte" varieties have been developed but the endophyte, a living fungus inside the grass, produces toxins which can lead to animal health problems. Biological control is available in the form of the parasitoid wasp Microctonus hyperodae. But Taranaki Regional Council expects that it will take 5-10 years for this parasitoid to build up and disperse to become an effective control agent of ASW. Assuming that 50% parasitism of ASW is then achieved, a 6% increase in pasture DM production could be expected (TRC Land Management Information Sheet No. 9). Contact the Regional Council or AgResearch for assistance.

Slugs

Slugs are a problem in wet conditions or where there is lots of trash. They are of particular concern to seeds and emerging seedlings of direct drilled crops because the drill slots provide favourable conditions for slugs. Problems with slugs can occur when numbers exceed 15 per m^2

Steps to take before sowing: Place 5 wet jute sacks diagonally across the paddock, three weeks before sowing. Cover the sacks with pegged down plastic to prevent them from drying out. If there is an average of more than nine slugs under each sack after two nights, slugs are present at levels that may damage the seedlings. Mob stocking at 500 ewes/ha for three consecutive nights prior to herbicide application can achieve good control of such slug populations.

Steps to take after direct drilling: Inspect drill slits two days after sowing. If there are more than four slugs per metre of row, then immediate use of slug baits is recommended, with the following options:

- Thiodicarb pellets @ 7.5 kg/ha if rain is imminent
- Methiocarb pellets @ 7 kg/ha immediately after rain
- Methaldehyde pellets @ 20 kg/ha immediately after rain.

The establishment of "beetle banks"

The establishment of "beetle banks" to maintain habitat for insect predators and to reduce the use of chemicals is recommended. When cultivating paddocks, create a ridge 0.4 m high and 1.5-2 m wide within the paddock. Leave a gap about 25 m at each end to allow farm machinery to move around it. These areas will accumulate vegetative material and provide habitat for insect predators, which will assist in keeping insect numbers down. As a guideline, a square 20 ha paddock (450 x 450 m) will only need one ridge in its centre to ensure adequate coverage by insect predators. The beetle banks will take two to three years to become effective forms of insect control.

1.5 STUBBLE RETENTION

Prudent farmers try to retain the stubble of the previous crop as a form of wind barrier, rather than burning the stubble, which is still practised by many. A research team at Crop & Food Research Ltd [Team leader Dr Prue Williams] is developing sustainable management systems for arable and horticultural lands and is studying amongst others what happens to the carbon in crop residues under various cultivation treatments [Glyn Francis et al.]. Crop & Food carried out a review on crop residues for the Foundation for Arable Research, Lincoln in 1996 (Dr Patricia Fraser, pers. comm.) but a copy is not yet to hand.

Evidence is mounting that all forms of cultivation oxidise soil carbon and recent US data suggest that the entire carbon content of a preceding wheat crop can be oxidised in 19 days following planting (Baker, 2000 b). Under regular cultivation, soil structure deteriorates, and thus the soil becomes more susceptible to surface and wind erosion. But with stubble retention and less cultivation, humus build-up occur at a rate depending on local soil and climate factors, and soil moisture is retained, especially in the top 300 mm of the profile, which leads to improved infiltration and crop yields (Baker 2000 b). He quotes Anon. (1994) who reported that when soil organic matter content is increased from 2% to 6%, a 65% increase in available soil water results. And that Triplett et al (1968) had found that no-tilled corn with 70% residue ground cover significantly out-yielded conventional corn, but with 45% residue cover, there was no difference between treatments.

1.6 STRIP OR CONTOUR CULTIVATION

In the USA, cropping lands prone to wind erosion are generally strip-cultivated, i.e. strips of vegetation, "grass hedges" are kept between areas that are cultivated and cropped. Those hedges act as major catchers of wind eroded soil, and of nutrients in surface runoff (see e.g. Eghball, Gilley, Kramer and Moorman 2000). Such "mini shelterbelts" have been evaluated in the Hawke's Bay by its Regional Council since the mid 90's (D Bloomer, pers. comm.), especially in asparagus growing regions. The HBRC brochure "Sustainable Land – Asparagus Strip Cropping" (May 1996) recommends strips of oats sown in mid June-July at 120 kg/ha over every second asparagus bed. The crop is sprayed off in end August at least 3 weeks before asparagus harvesting starts. In a season that lacked the typical strong Norwesters, the cost saving through lack of windblast to the asparagus spikes was valued at $750/ha. The total cost of establishing and spraying the oats hedges was $229. Asparagus pickers commented that harvesting conditions had improved as a result of increased shelter and minimal dust problems.

1.7 RIPPING OF WHEEL TRACKS

Wheel tracks are a key zone for initiation of surface runoff and erosion within paddocks. They are highly compacted, so have infiltration rates. They also collect runoff from higher ground and channel it. Various trials have demonstrated that cultivation or ripping of wheel tracks markedly increases infiltration and significantly reduces soil erosion.

Ripping compacted wheel tracks. ***Photo: Franklin Sustainability Project.***

Wheel track ripping, using a shallow tyned implement towed behind a tractor, should be carried out as soon as possible after planting.

1.8 COVER CROPS

A cover crop is grown to be ploughed into the soil rather than harvested at times of year when a paddock is not used to grow vegetables. It stabilises the soil against erosion, adds organic matter, improves soil structure, and uses residual nitrogen from previous crops. By improving soil structure, cover crops help retain high infiltration rates. Commonly used cover crops on granular loams include greenfeed or Massif oats, ryegrass, Phacelia and mustard.

1.9 FALLOW

At some stage it will be necessary to leave the soil fallow. The most effective measure to minimise erosion risk in this situation is to plough a cover crop back into the ground. The plant residues add dry matter to the soil, which helps bind it together. Other effective measures are to leave the soil in an open and loose condition, or form ridges using a tyned implement with only a few tynes.

1.10 SOIL STRUCTURE MAINTENANCE

1.10.1 Introduction

Retaining or improving soil aggregate structure and stability helps reduce surface erosion by sheetwash and windblow. Aggregate size and stability are important factors because a coarse structured, stable aggregate is relatively hard to mobilise but is quickly deposited when runoff/wind velocities decrease. A fine textured soil that easily breaks down to even finer sized particles is more easily mobilised by runoff or wind, and is more difficult to retain or settle out. Aggregate size influences soil transportability and is in turn influenced by aggregate stability. Both are affected by:

- Natural physics and chemistry of the soil type,
- Amount of organic matter in the soil,
- Type and frequency of cultivation.

The structure of most soils declines under cultivation and recovers under pasture. Soil structural degradation adversely affects cropping. Good soil structure leads to lower tillage costs, more days when the soil is suitable for cultivation, better seed germination, emergence, plant growth and vigour, and higher crop yields.

The type of cultivation equipment used can have a significant effect on soil structure. Machinery passage itself will physically alter soil structure. Rotary cultivators generally produce finer seedbeds than do ploughes, discs or tyned implements. The rate of change in structure stability is dependent on the soil type and on the cropping system. Long term, continuously cropped soils require more cultivation passes to produce a fine seedbed and are therefore, more susceptible to erosion, than are paddocks with short-term cropping, no-tillage cropping, or a pasture cover.

There is a strong relationship between aggregate stability and soil organic matter. Cultivation exposes more soil surfaces to aeration and this increases the decomposition of organic matter by soil organisms. This loss of organic matter contributes to loss of soil structure and reduces the ability of the soil to recover. Organic matter is conserved with minimum soil disturbance and maximum residue input, such as occurs under a no-till system.

1.10.2 General guidelines (from Shepherd et al 2000)

Minimum or no-tillage cultivation systems should be utilised. These avoid or minimise cultivation and minimise structural damage. Surface residue increases organic matter and therefore soil structure. Aggregate stability is significantly higher under these tillage systems than from cultivated cropping. Cultivation should be undertaken at the correct moisture levels. Cultivation when soils are too wet causes smearing and thin plough pans that reduce infiltration.

Cultivation implements that minimise structural damage should be selected. These avoid, or minimise, cultivation and therefore minimise structural damage.

Practices that avoid compaction of the soil should be used. These include using low ground-pressure machinery (wide tyres, tracked equipment etc); minimising traffic (by reducing cultivation runs; planning the paddock layout etc); and cultivating with a tyne behind the wheels to at least 10-15 cm depth.

- Ripping of subsoil pans should be carried out.
- Stock should be removed when the soil is too wet.
- Crop rotations that suit the soil should be used. Soils that are susceptible to compaction and structural degradation will require longer pasture and shorter crop cycles.

For further details readers are referred to the Visual Soil Assessment publications (Shephard et al, 2000), Volumes 1 and 2, Field Guide and Soil Management Guidelines for Cropping and Pastoral Grazing on Flat to Rolling Country.

Chapter - 2

Pasture Management for Surface Erosion Control

2.1 LOWLAND PASTURE MANAGEMENT FOR SURFACE EROSION CONTROL

Overgrazing of pasture can be completely avoided, though there are times in any year when feed is short and grazing pressure may be high. To minimise windblow, sheetwash or rilling, the critical thing is to maintain just enough ground cover, so that if a gale or a cloudburst strikes, a high proportion of eroded soil particles are trapped in the same paddock before they have a chance to move further.

Maintaining adequate ground cover fits in with things which most farmers do anyway in the course of grazing management: rotation of stock between paddocks, and avoidance of mob-stocking or heavy set-stocking during drought, cold conditions or wet weather.

Residual ground cover can be assessed during a paddock walk to determine if ground cover is adequate to protect soil from surface erosion;

- If one or two paces out of every ten strike bare soil or mud, ground cover is still enough to protect most of the bare patches against wind and rain.
- If between two and seven, cover is insufficient to prevent wind and rainsplash from detaching soil particles, but residual grass and litter still trap most of them before they move out of a paddock.
- If more than seven, detached soil is likely to leave in dustclouds or runoff during adverse weather.

The remedy is to spell the paddock if at all possible before it reaches this stage, and temporarily destock if it does. There is an obvious cost in terms of short-term feed utilisation. On the other hand, there is also faster recovery of pasture, and greater long-term production, if paddocks can be lightly grazed when weather conditions are adverse for plant growth. Measurements on farms show that maintaining sward density:

- Traps 90% or more of windblown and water-washed soil before it leaves the paddock;

- Helps reduce pugging and compaction of wet soil
- Prevents pasture growth losses of anything from 8 to 91% the following season.

The upper levels of pasture loss may seem high but have been recorded at diverse locations (Hawkes Bay, Manawatu, Canterbury, Otago). The highest losses occur where topsoil is stripped by water or wind; and when grass is trampled into mud on waterlogged, pugged soil. In either case growth cannot recover for a year or longer, even though climatic conditions have improved meanwhile. Light set-stocking or rotational grazing with adequate spells reduce the risk.

To ensure a healthy vegetative cover that will ensure erosion is minimised the following factors, that affect pasture/farm management, need to be recognised:

2.1.1 Climate

Temperature and day length are the two main environmental factors determining rate of plant production. Generally the higher the temperature, the greater the growth growth rates and higher the pasture production (as long as soil moisture and fertility are not limiting). Examples of this can be seen in NZ with winter growth rates varying from 2025kb dry matter/ha/day in temperate Northland to 5 kb DM/ha/day in Gore. The key is to select pasture species that are suited and adapted to the local climate and environment.

2.1.2 Pasture type and species

As a general rule where farmed animals are to be grazed the pasture ratio should be made up of approximately one third in legumes (ie red and white clover species) while the remaining two thirds in grasses (ie ryegrass, cocksfoot, timothy etc) and herbs (eg plaintain and chickory). The species will vary throughout NZ but should be selected as suitable for that particular environment. Where droughts are more frequent then one should include dryland type suitable grass/clover species which will withstand heavy grazing and periods without water. Some drought prone pasture mixes could include wheatgrass, cocksfoot, and fescue species as well as specialised lucerne pasture where suitable.

2.1.3 Soil fertility

New Zealand soils are naturally low in nutrient status, with phosphorus, nitrogen and sulphur being the most limiting macronutrients on unimproved soils. On newly developed pastures superphosphate and if necessary lime and molybdenum need to be applied to ensure the clovers are healthy and fix atmospheric nitrogen. As the soil nitrogen levels increase the grasses can establish and become more dominant thus ensuring a good effective ground cover for 12 months of the year.

The other important factor to consider is the replacement of nutrients removed by animals and lost from the soil. The soil type, stocking rate and level of nutrient reserves built up in the soil are very important factors in working out accurately the maintenance fertiliser rates. It is essential that an ongoing soil fertility monitoring programme is undertaken at regular intervals to ensure maximum pasture production (including persistence of plant species) and optimum animal performance is achieved.

2.1.4 Grazing management

From a soil conservation point of view the aim of grazing management is to ensure high pasture production as well as maintaining pasture quality which results is a complete protective cover over the soil mantel. Pasture regrowth follows the classic "S" shaped curve which has three phases where phase one is at the start after grazing with slow regrowth. If the paddock has been overgrazed (ie where less than 500kg DM/ha) then regrowth will be slow. With phase two we find the most rapid growth and phase three is a slowing down as more pasture length creates shading. Therefore if maximum pasture production was the only objective a system that kept the pasture in phase two would be best. Under rotational grazing this means grazing down to 35cm at each grazing and allowing them to grow back to 15-20cm before regrazing. But under a set stocking system this corresponds to setting a stocking rate so that pasture length is never below the height range of 3-6cm. However, maximum pasture production is not the sole aim in grazing management especially where we are protecting the soil from erosion. Pasture height may not be important compared to the spread of grasses and clovers (and other weeds and herbs) over the exposed surface.

Consideration of the types of grazing animals that are farmed is important because that will also determine the amount of pasture grazed and residual that should be left.

In summary the need is to vary grazing control and a stable vegetative cover is to have a sufficient number of paddocks which can only be achieved by temporary or permanent fences. This would necessitate having at least 30-35 paddocks on the farm and in the winter period mob rotational grazing (using temporary electric fences) can be used for the daily or tow day shift within a paddock.

2.1.5 Other factors

Pasture renovation

It is paramount that annual pasture renovation be carried out worked in with the farm rotation programme. As discussed elsewhere cultivation techniques may be desirable on some sites ie where there has developed a soil pan but from a soil conservation viewpoint where soil erosion is a problem and drought conditions prevail then it is strongly recommended that no tillage or direct drilling technologies be used.

Plant and animal pest control

It is imperative that ongoing maintenance programmes are carried out to control prevent or minimise any existing or new plant or animal pest.

Irrigation management

Using additional water from an irrigation system will ensure pasture species persist longer and help maintain the protective soil cover.

2.2 HILL PASTURE MANAGEMENT FOR SURFACE EROSION CONTROL

2.2.1 Introduction

This is the practice of continuously improving hill country pasture so that the ground is protected from sheet, wind, and shallow slip erosion by a close and vigorous mixture of clovers and ryegrasses. Subdivision fencing allows the control of grazing, thus improving sward density, and recent research has developed pasture species suitable for different climatic conditions and soil types.

2.2.2 Pastoral hill country

On hill country also, depletion of ground cover cannot be completely avoided during droughts or cold, wet winters when there is not much fresh growth. But it can be reduced by the same modifications to grazing management as for lowland pasture, i.e., maintaining greater residual feed in paddocks through rotational grazing, avoidance of mob-stocking or heavy set-stocking, and destocking the worst-affected paddocks (if feasible). There is also an additional technique which makes a big difference to density of ground cover on hill country – pasture improvement.

Since the 1940s, farmers have converted some 4 million hectares of hill country from low-producing to improved pasture, by oversowing and regular fertilisation. While carried out primarily to increase production, this practice has also had the beneficial side-effect of establishing a dense sward. Field trials have shown that establishment of improved pasture on hill country:

- Reduces surface erosion by 50 to 80%, relative to levels measured in unimproved pasture:
- Reverses pasture growth losses of 40% or more, (combined effects of overgrazing, sheet erosion and fertility depletion). In combination with closer subdivision and better grazing management, this reversal has enabled increases in stock carrying capacity from 2-7 stock units a hectare up to 8-12.

2.2.3 Utilisation of feed

It is important to utilise any extra grass feed to ensure pastures do not become rank (due to low stocking rates) and dominated by Cocksfoot, Yorkshire fog and lower fertility demanding or shade tolerant grasses and pests such as porina and grass grub. This utilisation can be achieved by more strategic fencing and increasing stock numbers over the periods when surplus dry matter is produced. Furthermore, aerial oversowing and topdressing will provide a better ground cover and more growth in the winter months. Clover response can be significant early on after oversowing until the available soil nitrogen levels are raised especially by specific topdressing and or rotational grazing.

2.2.4 Fertility transfer by stock

This can be one of the limiting factors in managing hill country pastures as stock tend to concentrate their grazing on the warmer crests of hills which can be overgrazed (bared off) and subject to frost heave, sheet and wind erosion. This important transfer of fertility results in varying nutrient levels and pasture production. The key is strategic fencing of shady blocks from the sunny so they can be grazed at the appropriate times with suitable stock types.

2.2.5 Natural reseeding

Where clover establishment has been patchy seed can be spread by stock following close grazing and summer spelling to encourage good seed set. The impermeable coat of clover seed enables it to survive through the animal and germinate in the dung. Some of the drier sunny blocks can be grazed in the spring and early summer then spelled to allow seed set and natural spread which will assist in improving the pasture ground cover.

2.2.6 Hieracium management

There are two opposing views on the role of hieracium species (hawkweed) in the hill country particularly unimproved lands which may support tussock grassland associations. One view sees the plant as beneficial because they are colonising land that can no longer sustain original grassland vegetation thus reducing the amount of bare ground which may be vulnerable to accelerated erosion. The other view sees these plants as aggressive invaders that have taken over from other species, so that their control would enable more desirable plants to be re-established.

There are four main options for controlling Hieracium namely, agricultural development, using herbicides, biological methods and grazing management. Grazing management appears to be one of the most feasible current strategies for limiting the spread of this plant onto low input land. Exclosure studies in Canterbury and Otago show grazing can reduce Hieracium flower density 40-fold limiting expansion by seed.

Low intensity spring-summer grazing significantly reduced plant number and ground cover of the upright Hieracium species ie praealtum (king devil) but not the prostate species ie pilosella (mouse ear). Conversely, high intensity grazing may assist establishment.

2.3 TUSSOCK MANAGEMENT FOR SURFACE EROSION CONTROL

Note that the comments and principles made above for hill pasture management also apply for tussock country particularly where the land lies below 900m a.s.l.

2.3.1 Introduction

Grazing management in the high country has changed dramatically over the past thirty years from traditional set stocking on large blocks of land to rotational grazing of small blocks at different times of the year. Furthermore, the period of grazing and stock grazing pressures are now recommended for each site.

There is no common set of recommendations for maintenance of improved tussock grassland. Each block, or group of blocks, will be managed for a specific purpose e.g. a sunny face to provide winter grazing.

There is now a greater tendency to manage each block individually. The basic objectives for successful pasture management and optimal dry matter as well as minimising the action of, sheet and rill erosion occur are to:

- Sustain a satisfactory level of production.
- Maintain the productive pasture species.
- Ensure good ground cover is maintained all year round.
- Ensure continuity of feed supply especially at critical stages of year.
- Utilise feed effectively so overall inputs are justified.

Level of production

Apart from soil moisture which can be manipulated, nutrient supply is the major factor affecting pasture production. It is well known that, in intensively managed pastures, maximum regrowth occurs when the sward is neither grazed too short (no less than 35cm), nor left to grow too long (no more than 12-18cm) before regrazing. This can only be achieved by mob stocking and rotational grazing. Intensive management is more difficult on tussock grasslands due to the terrain, unpredictable climate and most growth occurs during a short season. In such an environment it is more important to provide feedbanks where and when the stock need them than to achieve overall maximum pasture growth.

Maintaining productive pasture species

For sustainability and long term success from oversowing tussock land, it is vital to ensure pasture species persist and encourage spread of the introduced species. Because of the low soil nitrogen status, clovers dominate newly oversown pastures but as soil fertility levels improve grasses become more productive. There grasses have not been introduced annuals such as brome grass, browntop and sweet vernal will compete with the clovers. The practice of spelling for natural reseeding is more rewarding for grasses than for clovers.

Ensure good ground cover

It is essential to have an effective cover over the surface to minimise any erosion impacts. The most difficult time of the year will be in the winter and late summer when soil moisture deficits occur. Selection of suitable species which may include herbs for that environment is important.

Pasture quality when required

A high legume proportion in the sward, and a high ratio of green to dead material, will ensure good pasture quality especially when stock have high feed requirements such as ewes pre and post lambing and during flushing or mating.

Effective pasture utilisation

This can only be achieved by sufficient stock numbers and subdivision fencing. The aim is for 60-70% pasture utilisation on improved areas.

As a rule of thumb a stocking rate of at least one stock unit for every tonne of dry matter that is grown is required. For example, land in the semiarid areas which is moisture limited and producing approximately 2 tonne DM/ha/yr should carry at least 2 SU/ha/yr whereas a yellow brown earth soil in the mid tussock area producing 5 tonnes would carry approximately 5 SU/ha/yr. At least thirty hill blocks are needed to operate an effective management system that includes breeding replacement stock, and wintering on saved pasture. Because of the great variations in the mountain lands in New Zealand many different grazing practices occur. The approach taken here is to provide guidelines based on altitudinal, climate and general vegetation zones.

2.3.2 Guidelines

Alpine Zones

This zone varies around New Zealand, but basically occurs higher than 1500 m above sea level. In some localities it may include reasonable vegetative cover (alpine grasses and shrubs), but commonly has bare ground, rock and screes with snow cover for considerable periods. Frost heave is very active. Includes alpine vegetation.

Management Practices

Avoid unnecessary impacts on this zone. Where grazing is acceptable it should be at very low stock rates (eg 1 stock unit per 7-10 ha or more for short summer period i.e. 2-3 months). Care needs to be taken when setting stock rates as they type of season and environmental influences can impact on this. It is recommended in good growth seasons when there is a surplus of feed on the flats and lower hills of these properties that the summer alpine lands be spelled completely from grazing. Furthermore, this zone is suite to active and passive recreation so minimal tracking and soil disturbances should take place. Resource consents and approval from the Crowns agency will be required on any pastoral lease land for any development proposals.

- Recognise the long term impact of grazing on visual landscape. Most of these areas have important vistas and any overgrazing can be clearly seen at a distance. This applies particularly if any burning is carried out or escaped tussock fires occur. It is recommended that no burning be carried out in this zone.
- Where possible destock all grazing animals and remove feral pests. This will depend on the type of country and balance of the remainder of the farm. For it to be successful requires well sited retirement fences to keep farmed animals out of. It is important to note that in some of the more rugged terrain of the Southern Alps fencing may not be possible so the best alternative may be to graze solely cattle in the valley floor complexes.
- Where practical construct retirement fences. If Crown leasehold land, consider future land use options.

2.3.3 Retirement management practice

Practice

This is the practice of fencing out an area of severely eroded high country (usually LUC VII and VIII), which is often at the top of a mountain range. In some circumstances the topography of the area may enable stock to be completely removed without the need for large lengths of retirement fences.

Furthermore, to assist in the revegetation process of the retired land all stock that were removed will require alternative grazing elsewhere i.e. offsite feed. This is usually provided from the lower altitude, more productive land i.e. which may be AOS & TD or irrigated Lucerne to carry the same or equivalent displace stock on.

History

Retirement schemes were not really advanced at all until the Soil Conservation and Rivers Control Council provided grant incentives to farmers to carry out the practice from about 1955 onwards. This ceased in 1992 when grant monies for soil conservation purposes were terminated by government.

Application

Today the works still continues on private land but are at a slower rate unless it is part of the Crown Land Tenure Review process on Pastoral Leasehold land where the high altitude or severely eroded lands are retired and transferred to the Crown's estate to become the responsibility of DoC.

The same land use change can occur on Freehold land where severely eroding land exists and needs to be destocked to assist in the natural or man assisted revegetation programme. There may be opportunities to involve the Queen Elizabeth II Trust if there are unique areas of tussock grassland and bush within the same block.

The most important factor in any retirement block is to ensure that ongoing animals and plant pest control is carried out as well as regular fence maintenance. It is vital that proper fire management control such as firebreak tracks and watering facilities are in place in case any escaped or wild fires occur which will damage the vegetation recovery programme.

Tall Tussock Grassland Zone

This zone commonly in the 800-1500 m above sea level range is dominated by narrow leaved snow tussock (Chionochloa rigida) with sub-dominant blue tussock (Poa colensoi) and other inter-tussock species including herbs and weeds

There may be hawkweed ingression at lower altitude. Sunny faces are more depleted due to concentrated stock grazing. They are normally utilised for summer grazing in the December to early April, by dry sheep with stocking levels less than 0.2 stock units per hectare where unimproved. May include patches of native scrub and forest, particularly on the gullies and wetter sites.

- Use conservative stocking rates for the enterprise. This will depend on whether the land has been improved or not and how much sunny land compared to shady land exists. If unimproved land then stocking rates of 1 su/1.5ha-4.0ha may be acceptable (check local guidelines) whereas on improved land the rate can vary from 2 su to say 5su per ha per year on the best moist/warm country.
- Use conservative grazing approach.
- Aim for the best protective cover possible. From a soil and water conservation viewpoint this will require regular monitoring before stock are put out on a block and when to remove a mob from the block particularly in a dry summer period. Stock may need to be mustered from a block earlier than planned in a drought season to ensure an adequate ground cover is maintained into the winter. The recovery following a drought can be enhanced by applying seed and fertiliser in the autumn if required when sufficient moisture was available.
- Consider species diversity and develop only sustainable areas. This may mean very low grazing pressures on important biodiverse areas or alternatively fence them out and no grazing at all. Furthermore, areas should be developed according to the Land Use Capability classification of the land and consideration of the biodiversity values.

- Practice conservative burning policies (Refer Burning).
- Spell native blocks during good growth years. This is a valuable management tool particularly on blocks that have bare ground and may be unsuited to oversowing or further subdivision. This will help in seed set and may build up a small feed reserve.
- Use cautious approach to management of hieracium land.
- Continually monitor and check grazing loads. The fundamental principle from a soil conservation viewpoint is to ensure optimum ground cover at all times of the year. This means the blocks should be regularly monitored by eye or using various sampling/measuring techniques which will provide good data and trends on species composition, species change, amount of dry matter produced and bare ground recovery. This monitoring will provide information which will allow changes to be made at an early phase on the time and period of grazing, grazing intensities and type of stock used.

Short Tussock Grassland

This varies in altitude between 350 m – 800 m above sea level. In some locations it includes scabweed country, but the typical cover is depleted short tussock (Festuca spp) interspersed by exotic grasses and weeds. It is more typical of the South Island lower foothills and basin lands in the semi arid areas.

Some of this grassland was severely affected by rabbits and burning, but has recovered well with better fencing and stock control. Hieracium flatweed ingression is a major concern.

The supply of feed in this zone is critical to the viability of many run properties (i.e. early spring flush feed) and is oftenused for younger stock. The carrying capacity varies greatly depending on the seasonal rainfall and on farm development. Shady faces are vital for stock grazing in the late summer periods with average carrying capacity of 0.3-0.4 su/ha/yr.

Management Practices

- Constant management of rabbit and other animal pests as well as any major plant pest problems (including Hieracium spread invasion). The important factor here is for regular ongoing monitoring of the main animal and plant pest populations to ensure the threshold levels are not passed. Remedial plans need to be in place and ongoing control to ensure that a good ground cover is always maintained.
- Have drought strategies in place. Although droughts may be a regular occurrence on some of this country it is vital to have strategies in place to minimise the impact that lack of water and grazing animals can have on depleting the surface ground cover. Some of these matters haven been discussed previously such as deferring grazing or grazing only in the early

spring with the hogget flock and spelling to enable early seed set and natural spread of seed. The use of drought tolerant species on suitable paddocks at the lower altitude is advisable but grazing management and soil fertility status needs to be regularly monitored. Furthermore, using the most suitable lucerne pasture species for the site is recommended but special management is required (eg grazing and weed control) to ensure these valuable feed reserves persist for some time. Spelling lucerne blocks or deferred grazing may be necessary to restore them after a prolonged drought or heavy grazing. The other option is to have specific fodder feedbanks established at strategic sites around the farm which may or may not be used annually.

- Develop only sustainable areas that can withstand grazing. Otherwise consider managing them as forestry block or retire from grazing stock.
- Adopt conservative stocking policies. This is important on those lands which have been eroded and lost part or all of their topsoil as well as the threat of Hieracium plant invasion. It also depends whether they have been improved and the animal pest population threat.
- Aim for maximum protective cover. Similar management considerations to the other hill country zones.
- Exclude cattle from this land during wet periods. This is to minimise pasture species damage in improved areas but also will reduce the problem of cattle pulling out whole tussock plants (esp hard and silver tussock plants) in a wet season.
- Allow annual grasses to seed at least once per three years. This has been discussed elsewhere and is important to ensure improved and unimproved pastures persist. The provision of alternative feed sources as lucerne paddocks or special fodder banks and using irrigation on the lower flats and terraces will help facilitate removing stock off tender sunny country.
- Destock this land between November-February for seeding of sunny faces and spell blocks for a full season during good growth years. (Refer to comments above).

One of the most important management tools is to continually monitor and check grazing all the grazing loads.

Chapter - 3

Fencing Management for Surface Erosion Control

3.1 LOWLANDS

Fencing can also assist surface erosion control, if the fencelines follow the land types based on the principles of the NZLRI Land Use Capability Classification system. The key objective is to separate lower- producing, erosion prone units, from better arable or pastoral land.

Firstly, fence out Class VI and VII land and Class III & IV Land i.e. non-arable from arable. This practice can be used to minimise sheet and rill erosion on the steeper land by establishing more persistent pasture species. Trees can also be established on this steeper land, based on an open planting regime i.e. poplars/ willows, or on more vulnerable sites as a close planting regime i.e. pine or gum woodlots. Another alternative in the drier environments and sunny depleted faces, is fencing out these areas and establishing suitable "stock fodder banks" which may only be grazed once or twice a year. Suitable species will depend on locality but could include the following:

Shrub willows; Saltbush (Atriplex halimus), Dorycnium spp, Tagasaste-tree lucerne (Chamaecytisus palmensis), Tree medick (Medicago arborea), Shrubby wattles (Acacia spp), Perennial lupin (Lupinus spp) as well as nut trees i.e. Black locust (Robinia pseudoacacia), Honey locust (Gleditsia triacanthos), Carob (Ceratonia siliqua) and others such as Chestnut spp and various Oak spp.

Secondly, fence out "High Class soils" on the farm i.e. Class I & II from surrounding more erosion prone arable land. Better productive sites, without limitations of fragipan, soil compaction, poor drainage or stoniness can be intensively managed on a sustainable management programme.

Thirdly, fence out fertile alluvial floodplains and riparian zones (units with a w sub-class) from the adjoining landform. These flood-prone areas need special management such as no cultivation and preferably direct drilling for crop and pasture establishment, so as to avert sheet wash by floodwater.

3.2 HILL COUNTRY & MOUNTAIN LAND

Fencelines for hill country can also be made on the basis on LUC classifications. Separating classes I – V from the lower versatility units should be the first priority, where these classes are present on hill country properties principles for fencing these classes are similar to lowland areas.

Ideally class VI land should be fenced separately from class VII land as this will facilitate good grazing management, which will in turn reduce surface erosion. Class VIII should always be fenced off where possible for protection.

In the high country of New Zealand, fencing is still an important management tool for sustaining the soil and farm production. Although fences differ in design they have a similar objective, namely to facilitate stock grazing so as not to deplete the sensitive high country vegetative sward.

Fencing for the control of grazing animals has been practiced since the first sheep and cattle were introduced into New Zealand from the 1840s onwards. In the high country the basic design of fences has not changed other than more specialised wires i.e. 12.5 gauge, and using tanalised posts instead of flat standards where the site allows.

3.2.1 Size of flock and stocking rate

Size of the flock and block size is important if grazing sheep. Stocking rates will determine the paddock size ie 2000 ewes need 44 ha paddocks for grazing at 45 stock units per hectare if a proper rotational grazing system is to be practised.

3.2.2 Grazing Pressure

In general, there is more risk than benefit from periodic high grazing pressure on both unimproved and improved grasslands, especially in semiarid lands. On improved grasslands in sub-humid and particularly in humid zones, periodic high intensity stocking for herbage control and active nutrient cycling is essential as long as grazing is not to ground level.

3.2.3 Paddock shape and orientation

The key factors here are the area and length of fence and animal behaviour. The need to ensure stock water is available is paramount so watercourses should not be fenced out from stock in water short areas especially if a reticulation scheme is not available.

3.2.4 Soil Conservation Factors

Fencing will be to either act as an exclosure to stock (i.e. a debris avalanche) or as an enclosure to stock (i.e. to allow different grazing pressureson different land classes) e.g. sunny from shady. The latter may also be to spell plants for seeding or to permit seedlings to establish. The enclosure fence where possible, should follow the land-capability class boundaries wherever topography and climatic constraints permit i.e. snow drifts.

3.2.5 Siting factors

- Fences should be up and down a slope where possible to minimise damage from the snow
- Where slopes must be crossed look for stable sites, glacial ridges, old terraces and changes in slope
- For mustering purposes site fences on ridge tops so they can be seen (but not on exposed ridges as snow drifting may occur, preferable in this situation to be on the lee of the slope).
- Where acceptable from a soil and water and landscape perspective, dozing the line followed with regrassing may save costs in future.
- Cross streams at gorges or narrow confined points.
- Siting of gates is important for ease of access and movement of stock i.e. rotational grazing

3.3 SPECIFICATIONS

It is not the intention of this manual to list all the types and specifications as the range is almost unlimited with various combinations of HT (high tensile) and MS (mild steel) can be used. Boundary fences must be stockproof and built to last a long time.

The standard fence has seven wires, heavy strainers, posts and battens (or warratahs/metal standards) with posts 125mm or more about 4m apart. In the high country on areas subject to snow, steel standards, T irons and other sturdy material should be used. On the rough terrain standards/posts should be closer together with the wire strains being a lot shorter eg 200-240 meters. If electric fencing is to be used it should be of a standard that if the power is off the fence is still stockproof (Note electric fencing is not recommended for high altitude sites particularly retirement fences).

An alternative where snow drifting is a problem, is to construct specially designed fences that collapse in total (eg the "Hurricane chain fence") or where the wires loosen off. Nevertheless, maintenance of these snowline fences and other high country fences must be carried out on an annual basis to ensure they exclude stock from erodible areas.

Descriptions and illustrations follow for fences that are specifically to control grazing for the purpose of soil conservation.

3.3.1 Cattleproof fence

This is an existing fence (such as a summer snowline fence) upgraded to the standard of a new fence that will allow cattle to be grazed. It may involve wooden posts, metal standards and barbed wire or electric top wire. This enables cattle to be held within the block thus facilitating more even grazing of the tussock cover and reducing the need to burn.

3.3.2 Recuperative spelling fence

This is a fence erected on lower altitude country where pasture has been depleted by sheet, wind and frost heave. It is often dry sunny land that has a short growing season ie early spring flush with grass seeding later when the vegetative material dies out and exposes bare ground. Once a fence is constructed offsite feed provisions need to be planned so the stock can be completely removed from the block for a year or in the critical summer seed set time, enabling pasture recovery. Offsite feed may be from a fodder bank or irrigated pasture.

3.3.3 Erosion Control Subdivision or Conservation fence

These terms are synonymous with one another. The fences are used to separate eroded from non eroded areas, or to separate shady from sunny faces. They facilitate better grazing control, enabling eroded or sunny faces to be grazed for short periods, and stock to be moved of while pasture recovers.

3.3.4 Rabbit Proof fence

This type of fence can be made by placing rabbit netting (usually 40mm diagonal measure netting) over an existing fence, or constructing a new fence (on the property boundary or where rabbit numbers are high). It is essential to bury the bottom of the netting approx. 150mm into the ground, or if the surface is rocky then turn up on the ground and pack well down with stones to prevent rabbit burrowing. Gates should be swung close to a concrete or board sill buried in the gateway so that rabbits cannot wriggle underneath. Once erected, follow up poisoning, or using the RHD virus or other techniques such as shooting can be more effectively carried out to reduce the rabbit numbers and allow revegetation to take place.

3.3.5 Retirement fence

To specifically exclude all stock grazing from severely eroding land. This can take two forms. One is at high altitude, where severely eroded land is retired and left to naturally revegetate. This usually takes many years ie 15-20 before natural seeding and ground cover starts to show. Ongoing fence maintenance and feral animal control is essential for revegetation to succeed. The other form is on low altitude country that has been eroded. Here protection forestry species (for slope stabilisation and control of sedimentation) are established after all stock have been removed.

It is essential that all matters are considered when planning a retirement fenceline. Walk all lines before hand with an accurate measuring wheel noting all dips, angles, corners, where snow may lie and drift, gateway sites and creek crossings (floodgates, netting). Draw a profile of the line to assist in costings and type of material required. From the field notes a schedule of all the materials can be made and this is useful in pricing and tendering for materials and labour.

Rabbit-proof fence. *Photo: ORC.*

On the operational side, fencing gear may be best bundled and dropped by a helicopter. This will negate the need to construct unsightly access tracks, which may initiate further erosion. Each drop site should be well flagged and accessible for the fencing contractors. For additional information on specifications and design contact the local Regional Council. Any retirement fencing should be designed and constructed to last as long as possible, therefore the costs will be higher than other fences due to additional tiedowns, strainers and the like, being required.

Retirement fence, showing weed control as well as bush remnant.
Photo: John Douglas, Environment BOP.

Revegetation & Reversion

Chapter

Chapter - 4

Pasture Revegetation Practices

4.1 LOWLANDS

Lowland pasture revegetation may take place as a pasture renewal programme, following a cropping season at a whole paddock scale, or smaller filling in 'patches' from localised erosion. Lowland pastures are generally dominated by perennial ryegrass and white clover, other species can also be expected depending on the demands placed on the pasture by management and environmental factors.

4.1.1 Guidelines for lowland pasture revegetation

(Adapted from Langer, (1990), Pastures their ecology and management)

Timing

When water and nutrients are non-limiting, temperature plays a dominant role in the germination of herbage plants.

Traditionally new pastures are sown in the autumn. Autumn sowing allows the seed to go into a warm but sometimes dry seedbed, but with declining temperatures. These conditions are normally satisfactory for ryegrass, but species more sensitive to cool conditions may be disadvantaged.

Spring sowing will ensure adequate moisture and rising temperature, but the onset of early drought may cause some plant mortality where irrigation is not possible.

Fertiliser

Soil tests should be undertaken before determining fertiliser rates.

Phosphorus is essential for successful pasture establishment, particularly to stimulate the growth of seedling legumes. Generally applied at 20-35 kg P/ha when sowing, for some soils such as yellow-brown loams applications as high as 45-60 kg P/ha may be necessary. Superphosphate is often used, where sowing inoculated legumes, a reverted form should be considered as the strongly acid superphosphate may damage rhizobia.

Lime should be used to bring the pH of the soil to 6.0 or above to assist in the establishment of well-nodulated legumes, it is less important for grasses as they can tolerate lower pH levels.

Sulphur deficiency is widespread in NZ soils, but is usually corrected by 22-34 kg S/ha. Generally most deficiencies are overcome when superphosphate is applied at sowing. On soils such as brown-grey earths, where sulphur deficiency is much greater than that of phosphorus, sulphur-fortified super may be used. Elemental sulphur can be used where no phosphorus is required.

Potassium is often deficient on volcanic soils, although soils from greywacke are more commonly adequate in potassium. Legumes are less competitive than grasses in extracting available potassium from soil, and a deficiency can therefore result in poor legume establishment unless it is corrected by applying KCl or potassic superphosphate.

It is uncommon to apply nitrogen fertiliser when sowing new grass. The stimulus the grass component receives from the nitrogen may cause considerable competition with the accompanying clovers.

Seedbed preparation

When undertaking conventional cultivation the aim is to achieve a fine and firm seed-bed, so that the small grass and legume seeds make good contact with moist soil particles. Seeds should not be sown more deeply than 50mm, and frequently 25mm is more appropriate for small seeds.

Other guidelines for direct drilling, minimum tillage and conventional cultivation are covered in Chapter 1.

Species

A brief description of the most predominant pasture species is given here, seeding rates and mixes are not discussed. For more details on the suitability of different species, readers are referred to Burgess and Brock (1985), The Proceedings of the NZ Grassland Association.

Perennial ryegrass, generally dominates lowland pasture mixes. Generally rapidly establishing and requires high soil fertility, can struggle when faced with drought and insect damage. Several cultivars can withstand heavy treading and continuous grazing.

Italian ryegrass, although an annual, many types do persist for several years. Noted for their rapid establishment, which can jeopardise the establishment of other species. Requires high soil fertility and shows good growth in the autumn and winter.

Cocksfoot is well adapted to moderate fertility and low soil moisture. Some cultivars show tolerance to grass grub, and once gully established to argentine stem weevil.

Tall fescue, while slow to establish, does show a wide range of adaptability. Most tall fescue show reasonably good drought resistance and little frost damage during winter. Tolerant of acid and alkaline soils, withstand poor drainage, and some cultivars show resistance to grass grub and Argentine Stem Weevil attacks.

Phalaris is slow to establish, but once established withstands hard grazing and winter pugging and may persist for many years. Does well over cool seasons, but in dry summer conditions may become semi-dormant.

White clover, a wide range of cultivars are available with varying tolerance to severe grazing and treading. Generally requires soils of medium to high fertility, with evenly distributed rainfall. About 200 kg N/ha/yr can be expected to be fixed from the atmosphere.

Red clover, is easily recognised by its taproot, which aids in its tolerance to dry conditions. Requires high fertility and shows resistance to pests and diseases.

Grazing management

The first grazing of a new pasture is important and should be done before rapidly establishing species are able to shade out the less vigorous seedlings, the first grazing would normally take place six weeks following sowing when ryegrass is about 10cm high. Grazing should be done with a high concentration of light stock for a brief period. Frequent defoliation for the first 6-9 months will help the slower establishing species, especially legumes.

4.2 HILL COUNTRY

Slip oversowing involves the application of seed and fertiliser to bare ground left after the soil and sub soil has been removed by erosion. It is a temporary repair, pending spaced planting of trees or retirement from grazing to ensure permanent stabilisation.

The treatment of slip scars by oversowing dates back to the 1940s with work by Suckling and others at Te Awa in the Pohangina Valley in the Manawatu. (Suckling 1962) It was found that a number of factors lead to the success or failure. Available moisture was critical, followed by seed mix, fertiliser application and grazing control.

The advent of aerial sowing by fix wing aircraft, (McCaskill 1973) and later helicopters, led to larger operations in Hawkes Bay, Manawatu and Wairarapa regions after major storms. The costs of doing this work were high but in many cases subsidies reduced costs to farmers. Trails in the Whangaehu Valley in the Manawatu resulted in the restoration of slip scars to half of the production from uneroded areas, 7 to 8 years after a storm event (Garrett, 1981). Similar results were obtained in post Cyclone Bola oversowing trials in Hawkes Bay. (Quilter 1993) and (Lambert 1993) Although pasture growth will take decades to recover, and possible never reach pre erosion growth, research has shown that recovery time can be shortened by 20 years with the right management (HBRC 1999).

It was common practice during the 70s and 80s for subsidies to be made available through Catchment Boards for slip oversowing after an extreme storms. Oversowing was carried out from East Cape to the Wairarapa during this period.

The establishment of pasture on slip sites is usually very difficult because little soil remains and skeletal soils on eroded sites are subject to extremes of wetting and drying. Root development is restricted by thin or non existent soils and the plant's ability to obtain nutrients is very limited.

4.2.1 Guidelines for hill country pasture revegetation

Factors to be considered are:

Methods

Slip oversowing can be done by hand application, or for larger areas by aeroplane. A helicopter is the most efficient as many small areas can be treated at one time, especially if concentrated fertilisers are used, and the seed-fertiliser mix can be better targeted. Pelletised seed and fertiliser will aid germination and initial growth. Some farmers have placed a large mob of sheep on the area seeded for a short period to assist in bedding in the seed.

Moisture Levels

Treatment is recommended only when there is adequate soil moisture, or when there is a reasonable chance of rainfall occurring soon after the operation.

Seed Mixtures

Seed selections should include a mixture of legumes and grasses, which are tolerant of dry conditions and low fertility. Lambrechtsen (1987) recommends the mixture in Table 4.1. The clovers must be inoculated and all seed preferably pelleted.

Table 4.1 Seed mixtures

Kg per ha	Species
6-10	Ranui perennial ryegrass
4-6	Huia white clover
1-3	Mt Barker or Tallarook sub clover

Fertiliser Mixes and Rates

Fertiliser must be applied with the seed although there is some debate as to whether nitrogen needs to be applied if inoculated clovers are included in the seed mixture. Initial rates of 150 to 200 kg/ha are suggested, usually superphosphate with added minerals depending on the soil requirements. Diammonial phosphate at 100 to 150 kg/ha has been used with a repeat application 6 months later.

Grazing Control

The control of grazing by farm and feral animals is considered essential for the success of oversowing of slips. This can be difficult if large areas are treated at once, or there are large paddocks. Sheep will overgraze these sites, although they may also transfer some fertility.

A 12 months grazing restriction is the minimum suggested by most trials and larger scale operations.

The oversowing of slip scars on Class VII hill country has been carried out in small trials and larger scale programmes for over 50 years. Some of this work has been highly successful and has hastened the recovery of pasture production, while other examples have had minimal effect on recovery due to other factors such as weather, overgrazing and site exposure. At the best a 50% recovery after 5 to 10 years can be expected.

4.2.2 References

SUCKLING, F.E.T. (1962)., Te Awa Soil Conservation Experimental Area, Manawatu Catchment Board and Soil Conservation and Rivers Control Board, Wellington

HAWKES BAY REGIONAL COUNCIL, Environment Topics, Repairing Slip Damage 1999

Lambrechtsen, N.C. 1987 Grasses and Legumes for soil conservation, Plant Material Handbook.

4.3 MOUNTAIN LANDS

4.3.1 Introduction

Revegetation has been one of the most important erosion control methods used in the high country since the 1950s when topdressing and oversowing of bare depleted faces first started. The earlier techniques were basic but as technology improved e.g. helicopter application, using coated seeds with inoculum, and fertiliser, more suitable plant species, better ground cover has been established for longer periods.

When revegetation is carried out, alternative feed is provided to stock once they have been completely or temporarily removed from an eroded block. Offsite feed can be anywhere on the property but is normally on the lower altitude soils. Offsite feed can be pasture improvements, or supplementary feed, or an alternative fodder bank.

- The following practices can be used to revegetate eroded faces:
- Cultivation, direct drilling, minimum tillage

- Aerial Oversowing and Topdressing (AOS &TD)-pasture species
- Native plant establishment (may include hand planting)
- Forestry protection & production planting
- Dryland pasture & alternative crop production.

Historically there was a smaller range of plants to use in the 1940s and 1950s, but advanced research in plant selection, seed pelleting (with inoculum) and method of application has enabled better adapted plants for a variety of sites.

There is a multitude of research and publications highlighting various species for a particular site. The following guidelines identify key legumes, grasses, herbs and shrubs that can be safely used.

4.3.2 Guidelines for High Country Revegetation

- Follow the illustrated table for common species

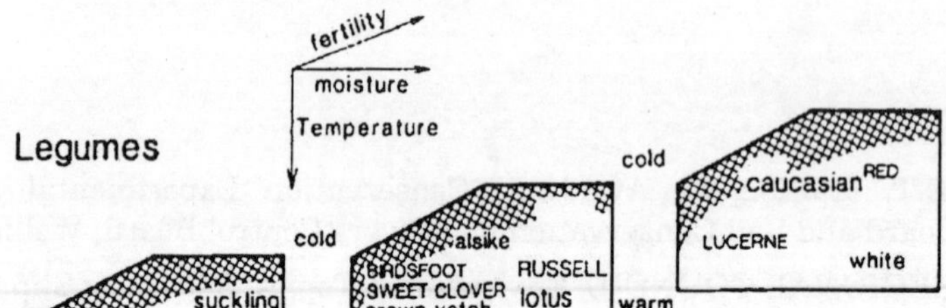

Figure 4.1 The most suitable pasture species in relation to temperature, soil moisture and levels of soil fertility. (After D. Scott and Keith Widdup, AgResearch, Lincoln)

- Legumes before grasses (select by species, and don't worry about cultivars)
- Most reasonable success is up to 800m above sea level (above this results are variable – depends on frost heave/shelter)
- Expect poor survival in harsh winters, be prepared to re-sow
- Well established species

Legumes

alsike clover

white clover

red clover

lucerne

Maku lotus

Grasses

cocksfoot

browntop/sweet vernal

chewings

timothy

perennial ryegrass

Yorkshire fog

- Species showing potential

Legumes

caucasian clover

perennial lupin

birdsfoot trefoil

canary

Grasses

tall fescue

tall oat

smooth brome

wheat grasses

Herbs & shrubs

salt bush

sheeps burnet

- Others which may have a specialised role

Legumes

zig-zag

crown vetch

sub clover

sweet clover

new trifoliums

Grasses

red fescue

phalaris

brome

- Seeding rate varies considerably, depending on the size of seeds and environment and can range from 6 – 10kg/ha.

Fertiliser Recommendations

Yellow- brown earth (brown soils): 250 kg/ha moly super in year one and 250 kg/ha Superphosphate every 2-3 yrs then lime as practical

Yellow-grey earths(Pallic soils): 200 kg/ha moly sulphur super (28% S) for 2-3 yrs Or 150 kg/ha moly sulphur super (33%), lime as reqd.

At high altitudes more fertiliser will be required due to the greater leaching rates but the economics will need to be considered. Soil tests are vital to ensure sunny and shady faces receive their own fertiliser maintenance recipe mix. Inaccuracy of fertiliser application is not good where bare ground is being revegetated, as stock can congregate on the fertilised strips and overgraze the new plants. Use precision equipment ie helicopter with high analysis materials. Avoid waste application on non-productive sites.

Chapter - 5

Sand Dune Stabilisation Practices

5.1 MARRAM PLANTING

History

As a primary sand stabiliser, marram grass (Ammophila arenaria) has been given most prominence in the past 80 years. Although it is a plant introduced from Europe, it has proven itself as a sand dune stabiliser throughout NZ. It is also found in some inland areas: on the Volcanic Plateau and at Erewhon Park in the Rangitata Valley. Since the early 1920s, it has been planted, more recently with mechanical planters, to stabilise sand dunes because it is easy to propagate and establish. With the increasing public concern about the 'natural character' of sand dunes, other plants are gaining more prominence. They will be discussed separately. The North American Ammophila breviligulata has been tried in NZ but was found to be less effective than the European A. arenaria.

Description

Marram grass (or European beach grass) is a tall, erect, spring and summer growing perennial grass, forming compact clumps with 60-80 cm high culms. It spreads rapidly through loose sand by strong rhizomes, several metres long, which send up new shoots through accumulated sand and with new roots forming at the nodes. The leaves are bluish green, up to 6 mm wide, ribbed on the upper side and sharply pointed. They are usually tightly rolled up to minimise transpiration. The flowerheads are compact, narrow, cylindrical spikelike panicles that develop from December to January. The seedheads when ripe are green yellow but marram grass does not set much seed under New Zealand conditions. It is vegetatively propagated by splitting two-year old plants into tufts.

5.1.1 Practice

Marram grass is always propagated vegetatively by transplanting bundles of culms gathered from vigorous 2- or 3 year old clumps.

Planting material can be collected from well-established existing stands, or from special selected sheltered "nursery" areas where the marram grass production is boosted by extra fertiliser applications of nitrolime (calcium ammonium nitrate) [or equivalent] at 125 kg/ha in spring and autumn. Or it can be obtained from some commercial nurseries, often by prior arrangement.

The planting stock is dug with a sharp spade 5-10 cm below the surface. The tillers should have at least 2 nodes and are trimmed to about 50-60 cm. Some 5 to 8 culms per bundle are planted in a 20-25 cm deep V-shaped hole and well heeled in. Spacing varies from 60 x 60 cm to 100 x 100 cm depending on the degree of wind exposure. Planting is in rows at right angles to the prevailing wind (preferably on the contour) while the bundles should be staggered in the rows to prevent wind funnels. Planting time is June, July and August. Small areas and steep dunes are planted by hand, while more extensive areas are machine-planted, using specially developed marram planters pulled by a crawler tractor, or a 4-wheel drive tractor with extra wide tyres.

Nitrolime (calcium ammonium nitrate) [or equivalent] applied at 45 kg/ha in spring and autumn will increase growth and early establishment.

Marram grass is more suitable for planting on top of the foredune, and on the main dunes and blowouts further away from the coast. It has also been successfully used to stabilise inland sand dunes near Cromwell and to control blowing pumice sand along the Desert Road, north of Waiouru.

Marram grass is unpalatable to stock.

Except for topdressing, the exclusion of stock, and the speedy repair of any blowout that may occur, no special management practices are required. Snails can be troublesome in newly planted areas and may need to be controlled.

5.1.2 Application

Where sand requires stabilisation with marram planting, the following steps should be taken first:

1 Contouring

Where possible, shape the sand dune to contours that blend in with the landscape and make the area less prone to the erosive forces of the wind. Any wind funnelling should be avoided and prevented. Where slopes are less than 15°, contouring can be done with a bulldozer, or possibly even with a tractor and blade.

Experience with repairing breaches in the foredune of major beach systems in New South Wales, Australia, following storm damage suggests that it is often more cost-effective to use heavy earthmoving machinery to reshape the beach and rebuild the foredune.

2 Temporary Retirement

The area has to be fenced beforehand to keep stock out. A standard 2-wire electric fence will keep out most cattle, so long as it is maintained and the power is kept on all the time. Since possums, hares and rabbits can still get into the area to be stabilised, they will need controlling also, unless there is good control in the adjoining areas.

3 Binding The Sand

On actively moving dunes in windswept areas, only a dense cover of marram grass will bind sand sufficiently for other plants to establish.

5.2 SPINIFEX PLANTING

Spinifex sericeus (formerly known as S. hirsutus Labill.) is a native of the New Zealand and Australian coasts where it occurs naturally on the seaward side of the foredune. It diminishes towards the south of the South Island and is not found much below Dunedin. With the increasing public concern about the 'natural character' of sand dunes, this grass is gaining more prominence and it is often used in preference to marram grass. The use of Spinifex requires long-term planning and commitment.

Spinifex has been described very comprehensively in the CDVN Technical Bulletin No. 2 "Spinifex on coastal dunes – Guidelines for seed collection,propagation and establishment" by David Bergin. See especially pages 21-25 "Recommendations".

5.2.1 Description

Spinifex is a 30 cm high stoloniferous perennial grass with stout overground creeping runners (stolons), several metres long which develop roots and leafy shoots at the nodes. The leaves are silvery hairy, 7-8 cm wide, with incurved margins and a fine tapered point.

It is one of the few dioecious grasses, the male and female flowers developing on different plants in the spring. The male flowers are arranged in loose terminal or axillary clusters. The female flowers develop into large 15 cm round, spiny terminal seedheads which when ripe in mid summer break off and are blown along the beach by the wind. Volunteer seedlings are often found amongst driftwood above high water mark.

Sometimes large colonies of plants of only one sex are found; these have grown from one single seedling.

5.2.2 Practice

Spinifex can be established from seed or from cuttings made from runners. Seed germination and strike rate of cuttings is often disappointing.

Seed is best sown in January-February when the seedheads can be collected from the beach, either as complete seedheads or as individual seeds removed from the seedheads. Seed should be buried 2.5-4 cm deep. Seed sown in spring gave much better survival and growth. Soaking the seed for 24 hours in seawater will improve germination. Seed can also be collected in mid to late summer and then sown in a nursery for planting out later, which may take 9-15 months. All seed should be free of smut fungus (Ustilago spinificis), which destroys the embryo. Or seed can be collected and stored in jute or paper sacks for sowing in the next spring.

Cuttings, particularly tip cuttings, 50-60 cm long, prepared from vigorous growing stolons are planted in early spring, some 20-30 cm deep and at a spacing of 1 x 1 m. For increased effectiveness, it is better to grow the cuttings first in a nursery with bottom heat and misting, and treatment with rooting hormones and a high phosphate starter fertiliser. Then a strike rate of 80% can be achieved, and plants will be ready for field planting in 18 months.

Spinifex can also be established by transplanting volunteer seedlings, carefully dug up without disturbing the root system, but the above nursery propagation techniques are more effective. They require long term planning and commitment.

Planting is at similar densities as those for marram: 50-60 cm in windy sites; up to 1 metre in more sheltered sites. At least 5 cm of sand should be above the potting mixture at planting. The addition of slow release fertiliser (e.g. Magamp) at about 30 g per planting hole and mixed in with the sand, assists plant establishment and vigour. Urea and other 'strong' fertilisers should be avoided.

Time of planting: spring or autumn planting gave similar results in the FRI trials, but summer planting should be avoided. Early spring planting [June to early August], well before the equinoctial gales, may be the best time because of more favourable temperature and moisture conditions.

Spinifex is less palatable than pingao, but may still be browsed by rabbits. Planted areas should be protected from disturbance by beach users through the erection of barriers and walkways; grazing animals should be excluded by fencing.

Maintenance should be by regular inspection of fences, walkways, and replanting where plants have died. Without a uniform vegetative cover, wind may dislodge sand again. Fertiliser applications on established stands at 200 kgN/ha increased cover by up to 52%. One application remained effective for 3 years.

5.2.3 Application

Spinifex is mainly used for foredune planting, especially to repair gaps in the vegetative cover on the seaward side. For example, near walkways and other high use areas, where closer planting may be desirable. Without disturbance, the extensive stolons grow down slopes and in hollows collecting sand and smoothing out the duneface.

Because of the need for nursery propagation of Spinifex for its effective establishment, the plant lends itself to community involvement, such as Beachcare or Coastcare groups. It is less readily applied to large-scale revegetation projects, unlike marram grass. The preparatory steps outlined for marram, apply here also, with possibly greater emphasis on local community involvement.

1 Contouring

Since Spinifex is more commonly used on the foredunes, it is even more important to shape the sand dune to contours that blend in with the landscape and make the area less prone to the erosive forces of the wind. Any wind funnelling should be avoided and prevented. Where slopes are less than 15°, contouring can be done with a bulldozer, or possibly even with a tractor and blade.

Experience with repairing breaches in the foredune of major beach systems in New South Wales, Australia, following storm damage suggests that it is often more cost-effective to use heavy earthmoving machinery to reshape the beach and rebuild the foredune than using manual labour.

2 Temporary Retirement

The area has to be fenced beforehand to keep people and stock out. A standard 2-wire electric fence will keep out most cattle, so long as it is maintained and the power is kept on all the time. Timber railings and timbered walkways will be necessary to control pedestrian traffic. Since possums, hares and rabbits can still get into the area to be stabilised, they will need controlling also, unless there is good control in the adjoining areas.

3 Binding The Sand

On actively moving dunes in seaward areas, only a dense cover of Spinifex, usually with pingao, will bind sand sufficiently for other plants to establish.

5.3 PINGAO PLANTING

History

Desmoschoenus spiralis (pingao) is a sedge that occurs only in New Zealand where it can be an effective sand binder on foredunes, but it is much more difficult to propagate, so that it has not considered as a major stabilisation species. It has always been valued in Maori weaving for its yellow colour. With the increasing public concern about the 'natural character' of sand dunes, this sedge is gaining more prominence and it is often used in combination with Spinifex and in preference to marram grass to stabilise foredunes and other seaward-facing situations. The use of pingao requires long-term planning and commitment.

Recent research by FRI has shown that pingao can be used as a sand stabiliser. See CDVN Technical Bulletin No. 1 "Pingao on coastal sand dunes" by D O Bergin & J W Herbert for detailed information; especially pages 18-19 "Recommendations".

5.3.1 Description

Pingao, also called golden sand sedge, is endemic in NZ, the only representative of its genus. It forms rope-like, rather woody rhizomes on foredunes; its tufts of coarse, grass-like curved leaves are 2-5 mm wide, 30-90 cm long, smooth and varying in colour from green to orange; its culms are triangular in cross-section and 30-90 cm long. The seedheads are panicles, which carry nut-like seeds that are smooth and dark brown. The seeds are arranged in a spiral pattern.

Pingao was common throughout NZ but always rare in Southland. Nowadays, there are only scattered populations of pingao, due to grazing and browsing animals, and competition with marram. It is often associated with Spinifex, which grows in similar situations.

Vigorous stands of pingao are observed where there is sand movement as on the seaward faces of foredunes. Young plants can withstand 10-20 cm of sand accumulation, whereas well-established stands can tolerate up to 70 cm of sand annually. But excessive accumulation or erosion causes dieback. Esler found on the Manawatu coast that pingao was less resistant to sand removal than other sand binders [Spinifex and marram], although its woody rhizomes may persist.

FRI studies have shown that there are distinct provenances of pingao, some with a sprawling rhizomatous habit, and others with a more tussock-like form. Others differed markedly in frost-tolerance and important differences in weaving properties were also found. It is therefore important that local material be used in revegetation plantings.

5.3.2 Practice

Pingao is best established from seedlings. Planting from rhizome cuttings is much less reliable. Seed can be collected in large quantities during December to early January, depending on summer temperatures. The seed should be shiny brown-black when collected. It should be stored in jute or paper bags, not plastic bags. The seed germinates readily with husks still attached. Seedlings should be 9-15 months old by the time they are planted in autumn or very early spring.

Time of planting: some people prefer autumn planting; others winter or early spring planting. In one trial, spring planting resulted in a much higher survival rate of pingao. At planting, the plants should be 25-30 cm tall and have at least 5 leaves. As for Spinifex, they should be planted with at least 5 cm of sand above the potting mixture; 30 g of slow-release fertiliser (e.g. Magamp) should be mixed in with the sand in the planting hole. Planting densities as for Spinifex and marram: 50-60 cm in windy sites; up to 1 m apart in more sheltered sites, and when using the rhizomatous strains.

Maintenance should be by regular inspection of fences, walkways, and replanting where plants have died. Without a uniform vegetative cover, wind may dislodge sand again. Fertiliser applications on established stands at 200 kgN/ha increased cover markedly. Since browsing by rabbits and other animals is a major cause of pingao failure, control of rabbits and hares is essential, as is the exclusion of grazing animals, such as cattle, sheep, goats and deer. The planting sites should also be protected from disturbance by beach users. Suitable barriers and walkways should be erected.

5.3.3 Application

Just like Spinifex, pingao is mainly used for foredune planting, especially in clusters to simulate its natural growth patterns. Since excessive sand accumulation or erosion will retard growth or cause dieback, survival and growth of pingao can never be guaranteed. Where strong on-shore winds cause high rates of sand movement, pingao should not be used. Marram would be more suitable. While marram could be used as a "nurse plant" for pingao, there is evidence that it could out-compete pingao, particularly on exposed sites. The use of pingao should therefore be more carefully considered beforehand than that of Spinifex and marram.

Because of the need for nursery propagation of pingao for its effective establishment, the plant lends itself to community involvement, such as Beachcare or Coastcare groups. It is less readily applied to large-scale revegetation projects, unlike marram grass. The preparatory steps outlined for marram, apply here also, with possibly greater emphasis on local community, including iwi involvement, if the pingao is to be used for weaving once well-established.

1 Contouring

Since pingao is more commonly used on the foredunes, it is even more important to shape the sand dune to contours that blend in with the landscape and make the area less prone to the erosive forces of the wind. Any wind funnelling should be avoided and prevented. Reshaping with machinery should be considered if manual methods are likely to be ineffective or too costly. Temporary shelter in the form of straw/hay bales and shelter cloth may be appropriate, if the likelihood of successful pingao establishment is high.

2 Temporary Retirement

The area has to be fenced beforehand to keep people and stock out. A standard 2-wire electric fence will keep out most cattle, so long as it is maintained and the power is kept on all the time. Timber railings and timbered walkways will be necessary to control pedestrian traffic. Since possums, hares and rabbits can still get into the area to be stabilised, they will need controlling also, unless there is already good control in the adjoining areas.

3 Binding The Sand

On actively moving dunes in seaward areas, only a dense cover of pingao, usually with Spinifex, will bind sand sufficiently for other plants to establish.

Note:

The CDVN Technical Bulletin No. 3 "Sand tussock on coastal dunes – Guidelines for seed collection, propagation and establishment" by David Bergin, describes the use of Sand Tussock in great detail. But since this grass "plays a minor role in sand stabilisation and foredune development", it has been decided not to include it in these "Soil Conservation Practices". Readers who wish to use this grass are advised to consult the Technical Bulletin.

5.4 COASTAL FOREST AND PASTURE PLANTING

Stable sand dunes were covered with native shrubs and trees before the arrival of man. Few areas are left with relatively undisturbed coastal forest; many areas have been converted to a planted forest of Pinus radiata, usually with a belt of macrocarpa facing the seawinds. Increasingly, local residents want to convert coastal areas to the vegetation that once was there. To determine the composition of that vegetation requires local expertise or knowledge that should guide planting options and strategies. Long-term planning and commitment is essential for those locals who wish to embark on this path with assistance from local/regional councils and scientific advisers.

5.4.1 Description

This section will deal with a selection of native and introduced plants that are known to be tolerant to coastal conditions, and with a planting strategy that may bring success. A prerequisite is that the area to be planted has already been stabilised with the grasses/sedge described in the previous sections.

5.4.2 Practice

Once the dune has been stabilised, planting a permanent ground cover of shrubby or even tree species can be attempted. These species need to be hardy and persistent, and preferably nitrogen fixing. Yellow tree lupin (Lupinus arboreus) was commonly used but it has been affected by the fungus Colletotrichum gloeosporioides in recent years, so that it is less common today. But it appears to be coping with the blight, and its use should be reconsidered. Alternative shrubby legumes, such as tree lucerne (Chamaecytisus palmensis) and some shrubby wattles (Acacia sophorae and A. cyanophylla) are known to survive in coastal areas. The latter are currently not recommended because they spread by suckering and seeding, and can suppress other vegetation. And tree lucerne requires reasonably sheltered locations.

Whether one wishes to use these introduced legumes to enrich the sandy environment with nitrogen prior to planting shrubs or trees, is a local decision, dependent on local regulations, if any. The alternative is to use slow-release NPK fertilisers at planting, as described in the previous sections.

1 Native Species

Native species may be planted for preference, or for conservation areas, or to complement areas set aside as reserves. The following native species are wind and salt tolerant, and will eventually form a dense stand. They can form the nucleus of a coastal forest and have been ranked in order of preference.

Phormium tenax	Flax
Coprosma repens	Taupata
Pittosporum crassifolium	Karo
Pittosporum tenuifolium	Kohuhu
Griselinia littoralis	Papauma or Kapuka
Myoporum laetum	Ngaio
Metrosideros excelsa	Pohutukawa
Muehlenbeckia complexa	Pohuehue
Cortaderia fulvida / toetoe	Toetoe

Other native species are harder to establish on sand dunes, or may not be found in the particular botanical district. Advice should then be sought from the Dept of Conservation, and/or land management officers of regional councils.

2 Permanent Trees

As an alternative to the use of native species, a permanent tree cover can be planted to provide shelter, or a tree crop, or a mixture in the form of agroforestry. In general, a mixture of native and introduced species is recommended with the choice of species rather dependent on what the land manager can afford, and desires as end use of the trees.

3 Permanent Pasture

To establish and maintain permanent pasture on sandy soils can be very difficult because such soils usually have these features:

- Low natural fertility
- Inability to hold moisture
- Lack of organic matter
- Proneness to erosion

These features, together with exposure to strong, often very cold salt-laden winds, dry summer conditions, and often significant areas of steep contour, mean that special management techniques are needed to establish and maintain a pasture cover that will provide sustainable stock grazing and avert the risk of wind erosion. When pasture has been depleted, sand will start to blow again.

5.4.3 Application

1 Native Plants

The native plants, preferably obtained from local seed sources, should be hardened off one month prior to planting between May and August at 2-3 metre spacing among the grass/sedge vegetation which has effectively stilled sand movement. The earlier they are planted, the better. The addition of fertilisers such as nitrolime and magamp will help establishment and survival. Controlling weeds around the plants in their first year will also help to ensure survival until they are big enough to suppress weeds. Kikuyu grass may have to sprayed carefully with a herbicide such as "Gallant" or equivalent; it tends to smother plants otherwise.

Since native plants are quite palatable to stock, an effective permanent fence is a must.

2 Permanent Trees

Shelterbelts

For successful plantings, a permanent fence is a pre-requisite. The area should be about 7 metres wide for best results with an appropriate mixture of plants, at right angles to the prevailing wind. The narrower the area, the less effective the shelter, and the more likely that stock, especially cattle and horses, will reach over the fence and damage the plants.

During May to August, plant a sequence of flax (Phormium tenax), pohutukawa (Metrosideros excelsa), taupata (Coprosma repens) or ngaio (Myoporum laetum), and macrocarpa (Cupressus macrocarpa), with flax facing the sea winds.

Woodlots

The hardier timber species, used for farm forestry, can also be considered in coastal areas when they are salt wind tolerant:

Pinus radiata	Radiata pine
Pinus pinaster	Maritime pine
Pinus muricata	Bishop pine
Pinus nigra	Corsican pine
Cupressus macrocarpa	Macrocarpa or Monterey cypress

Cupressus leylandii	Leyland cypress
Eucalyptus botryoides	Southern mahogany gum

Cypresses and pines usually achieve better growth rates and forms in a coastal environment. The growth of gums and wattles can be highly variable, depending on site conditions. Silver poplar and some shrubby willows can be used to provide shelter for more productive tree crops. Other poplar and willow species are not recommended because of poor growth form and susceptibility to salt spray. From 100-500 metres landward from the foredune or cliffs, spray burn and wind shear will stunt timber species. But beyond 500 meters, normal growth rates and forms can usually be obtained behind the shelter provided by the trees to windward.

In the autumn prior to planting, establish suitable planting sites in the existing vegetation. Do not clear that vegetation if a new phase of wind erosion is to be avoided. Plant tree seedlings at 1000 stems per hectare (3metre spacings) between June and August.

For management, do not thin or prune trees in the zone 100-200 meters landwards from fore dune or cliff; maintain as a protective shelterbelt. Between 200 and 500 metres, a standard pruning and thinning regime can be applied to pines and cypresses. Prune up to 6 metres in successive lifts, and thin to a final stand density at 200-400 stems per hectare.

Agroforestry

At sites more than 500 metres from fore dune or cliff, agroforestry can be considered, preferably if a protective planting, or at least an adequate shelterbelt has been established in the zone from 100-500 metres. Any of the timber species recommended for woodlots can be used for agroforestry. Plant tree seedlings at 200-400 stems per hectare in June to August in sites pre-prepared during the autumn.

Sheep have to be excluded for 1-2 years; cattle for 3-4 years until the seedlings are sufficiently high to escape browsing. If this is impossible, the seedlings will have to be individually protected to establish the agroforestry block that will provide highly desirable shelter for stock.

While standard pruning and thinning procedures can be followed, the risk of creating 'top heavy' trees should be watched carefully in this windy, coastal environment. Top heavy trees have been much more prone to breakage when thinned to 50-100 stems per hectare.

On sandy soils, the loss of pasture production as the trees get heavier crowns, is likely to be greater than that recorded on heavier soils. But the beneficial shelter effect to stock is also likely to be greater in this coastal environment with its strong, often cold sea winds.

3 Permanent Pasture

To establish pasture on sandy soils, the land should firstly be of an even contour. Shaping with bulldozer or tractor with blade to prevent wind funnelling may be required beforehand. Then a cereal, such as barley or oats, or annual lupin, which establishes quickly, should be sown as a covercrop in autumn, with the pasture species germinating more slowly under the cereals. Careful grazing will be required to maintain the covercrop, while allowing the perennial pasture species to come through.

To revegetate small areas such as blowouts, see sections on Marram, Spinifex or Pingao.

To maintain pasture on sandy soils, the key components are:

- Soil fertility
- Pasture species
- Pest control
- Grazing management
- Choice of stock

Soil fertility

Sand country is naturally low in organic matter and the major plant nutrients, NPK. The free draining nature of these sandy soils causes rapid leaching of fertiliser. While improvements in soil fertility and organic matter can be achieved over time, careful management is required to maintain them. To improve soil fertility, an appropriate fertiliser programme needs to be determined by

- Getting the soil and stock health tested to determine their current status
- Assessing nutrient losses from the farm by determining current stocking rate, stock type and management, pasture yield and management, and soil type
- Based on the above, working out fertiliser application rates for annual maintenance and improvement of soil fertility status.

Because leaching of nutrients on sand country can be high so that it is difficult to build up the nutrient status of the soil, the following practices should be considered:

- Slow release fertilisers e.g. RPR, especially on acid soils
- Organic fertilisers e.g. poultry manure to build up organic matter and moisture and nutrient retention
- Potassium applications in the spring
- Nitrogen applications to coincide with pasture growth e.g. spring and autumn
- Fertiliser applications little and often.

Pasture species

Kikuyu grass *(Pennisetum clandestinum)* is extremely common in dune regions from Waikato and Bay of Plenty northwards, but may be found in dunes further south. While it is frost-tender, it did persist for 2 winters at Lincoln. As a pasture grass, it needs heavy grazing in late spring-early summer to maintain palatability. On sand country, it can die off in severe summer droughts and become a fire hazard, or it can lead to gaps in the sward that can then lead to blowouts. If not grazed hard, it can be very competitive with existing vegetation, and seedlings of native shrubs and trees have very little chance of establishment among kikuyu (Edgar & Connor 2000, p. 575). On the other hand, its mat-forming growth habit makes it a suitable species for erosion control. Where kikuyu dominates pastures, efforts can be made to introduce and maintain temperate grasses to obtain better autumn and winter production. The following pasture plants can complement kikuyu:

Temperate grasses

Perennial ryegrass varieties	*Lolium perenne*
Annual ryegrass varieties	*Lolium multiflorum*
Cocksfoot	*Dactylis glomerata*
Prairie grass	*Bromus willdenowii (B. unioloides)*
Yorkshire fog	*Holcus lanatus*
Tall fescue	*Festuca arundinacea*
Phalaris	*Phalaris tuberosa*
Legumes	
White clover	*Trifolium repens*
Red clover	*Trifolium pratense*
Lucerne	*Medicago sativa*
Subterranean clover	*Trifolium subterraneum*
Birdsfoot trefoil	*Lotus corniculatus*
Herbs	
Plantain	*Plantago major or P. lanceolata*
Chicory	*Cichorium intybus*

These species are best introduced by undersowing into a kikuyu sward that has been suppressed by heavy grazing, harrowing or spraying. In the absence of a kikuyu sward, these species can be undersown into a covercrop such as barley or oats, or annual lupins.

Pest control

Not only do pests such as hares, rabbits and possums have to be controlled in dune country if pasture is to establish and persist, but also slugs and snails, and various insect pests, such as grass grub, black beetle, black field cricket, and clover flea have to be controlled. Plant pests, such as thistles and ragwort can establish in pastures, opened up by insect damage.

Grazing management

Because of the difficult and changeable conditions prevailing in coastal sand country, grazing management has to be careful and flexible to cope with these conditions that can lead to surpluses and severe deficiencies in food availability. A rotational grazing system should be designed around:

- Adequate subdivision
- Adequate stock water and shelter
- Deferred grazing and spelling

Any incipient sand blows should be fenced off and treated immediately with grass/sedge planting (see sections on Marram, Spinifex or Pingao).

Choice of grazing stock

The choice of stock for sand country can have a major impact on pasture persistence, or the recurrence of blowouts, and should be carefully considered. In general, cattle damage coastal pastures far more quickly than sheep, but light-weight cattle breeds with carefully controlled grazing regimes may keep such damage to acceptable levels that prevent recurrence of erosion.

Chapter - 6

Burning Management – Mountain Lands

6.1 INTRODUCTION

In the high country of New Zealand burning has been carried out for many centuries. From the 1840s it was used to open up rank tussock grassland, which was unable to be readily grazed by sheep. Without any controls, escaped fires frequently burned several hundred hectares of erosion prone land. Stock were commonly grazed back on the burnt areas, overgrazing any regrowth and camping on the sunny faces, thus accentuating frost heave, sheet, and wind erosion.

It is now recognised that burning of vegetation should only be carried out in special circumstances and not used as a regular development tool each year. Alternative options may have minimal impacts on the soil and water. Burning on large blocks is difficult to control and fraught with many risks if the wind changes.

Before deciding to burn, consider:

- The objectives for burning tussock and intertussock grasslands (land clearance, rejuvenation of native pasture or removal of unwanted shrubs or weeds);
- What conservative burning practices are appropriate given the various land types and vegetation being burnt;
- Time since the previous burn. Burnings impact on vegetation depends on the history of management (burning of snow tussock should be only carried out once every 12-15yrs. to avoid depletion of biomass)
- Post fire management. It is essential to burn when the ground is damp in the winter months and spell the site afterwards. Follow up AOS &TD where feasible on the lower altitude sites will be beneficial. Lenient grazing in the first two years following the burn, will assist regrowth. Ensure rabbits and hares are controlled on the sunny faces.
- Whether post-fire spread of ingressive weeds e.g. hieracium spp., can be controlled
- In the wetter Lakes regions where fern and scrub are a problem burning may need to be carried out in late winter followed up with AOS & TD and erection of erosion control fence(s) for ongoing grazing control.

- Have a Burning Management Plan in place for a five year period (incorporate with the Farm Plan programme).

6.2 GUIDELINES FOR GOOD BURNING

Source: Otago Regional Council Draft Code of Practice for Vegetation Burning in Otago 2000.

Safety and control (Risk Management)

- Ensure that the necessary consents or permits are obtained from the relevant agencies and be familiar with your insurance policy provisions.
- Seek advice from a meteorological service prior to burning on the expected weather conditions.
- Avoid lighting fires while a strong wind is blowing, when a strong wind is predicted, or when conditions are such that a fire is likely to spread beyond the limits of the area that is intended to be burned.
- Use experienced people to light and control a fire.
- Do not leave fires unattended.
- Be familiar with the contact phone number of the local rural fire officer for the rural fire authority.
- Ensure that a helicopter is available if it is needed.
- Where available, use a snow cap to control the area to be burned and form an effective barrier for fires.
- Make use of fire breaks where a snow cap is not present by pre-burning prior to the main burn, to prevent the spread of fire into unwanted areas.

Controlling the Nuisance Effects of Smoke

- Endeavour to burn when weather conditions are such that smoke is likely to dissipate by the following morning.
- Avoid conditions that could result in ash deposition on skifields during the ski season and other recreational uses.

Firebreaks and access tracks

Firebreak tracks which can be dual purpose for vehicle access and stock movement have been used for many years. The main objective is to act as a defined structure where a fire can be lit from or alternatively be used as a back burning barrier. The other role is to form a normal control barrier for any natural or escaped fire thus allowing men and vehicles onto the site to help control the burns.

Since about 1950 a large number of firebreak tracks were constructed using grant assistance. In the South Island emphasis was on comprehensive range and mountain land "Strategic Firebreak Access Schemes" where all properties were linked into an overall community fire control management plan. This worked well where one had the cooperation of landowners. Within properties was another matter.

Summary of Guidelines

- Consult with the appropriate rural fire authority
- Plan the over all route proposed using local knowledge and expertise e.g. geologist if potentially unstable slopes are present. Survey the whole track and mark all strategic corners, where culverts or cutoffs should be with pegs or coloured cloth so the operators can see from a distance
- Site so there is minimal landscape disruption
- If feasible, use a hydraulic excavator in preference to a bulldozer, to minimise site disturbance
- Use only experienced operators with the appropriate horsepower machinery
- In some instances an environmental impact assessment may be advisable (check on Resource consent requirements)
- The average width of the carriageway should be 3 metres
- The grade of the track should not exceed 1:8 (i.e. 7 degrees) except where ground conditions demand eg hard rock, bluffs. Here for short distances, grade can be increased to 1:5 (i.e. 11 degrees)
- On hill formation, frequent culverts or cutoffs need to be constructed with rockwork at both the inlet and outlet to prevent erosion of water tables, downhill batters and waterways.
- Stream crossings should be constructed to a hard bottom. Alternatively they can be armoured with rockwork, rock filled or adequate sized culverts placed (i.e. able to take storm flows)
- Uphill batters should be graded back to a slope sufficient to minimise slumping and regressing.
- Revegetation of the disturbed land should be carried out as soon as practical with suitable species according to the site and altitude. A general mix could include 3kg White clover, 2kg Montgomery Red or Alsike clover, 3kg Yorkshire fog, 3kg Browntop/Fescue and 2 kg Cocksfoot plus suitable fertiliser as 100kg Sulphate of Ammonia and 300kg Molybdic superphosphate (or sulphur). Apply the dressing with a tractor mounted spinner or blower or helicopter bucket.

Chapter - 7

Fodder Bank Establishment

7.1 INTRODUCTION

This practice has been around for some time but it is only recently that special species have been planted as alternative fodder. The technique is applicable to drier environments in the high country, but may be useful anywhere in New Zealand. It involves establishing specially adapted plants that will provide a bank of additional supplementary feed on one or two small areas, once or twice a year. Fodder banks may be planted on sites subject to frost heave, sheet and wind erosion but in general are established for "offsite feed at low altitude" to enable higher altitude eroded sites to be destocked or permanently retired.

Fodder banks have become more prominent due to many successful farm trials and research plots in particular in the 1980s in Marlborough, the Hakataramea Valley and Central Otago. These provided useful data on growth form, dry matter production and the best management practices. Traditionally fodder banks have been used as a drought management strategy, but today they are considered a wise option in erosion control.

7.2 GUIDELINES FOR FODDER BANK ESTABLISHMENT

Because of the large number of plants available to use in fodder banks general guidelines only will be provided. More specific details should be sought Agresearch, Alexandra. The establishment of any feedbanks should be undertaken as an integral part of the management of the whole property.

The key to these new species is that they are suited to harsh sites, often dry and of low fertility. Many of the earlier species were grasses and legumes, now the trend is to use shrubs and other herbs as well.

The success of fodder banks depends on planning, in advance so all the details on the soils, fertility, moisture regime, area, type of fencing, type of grazing stock suitable, planting equipment and where to purchase large numbers of quality plants, are known.

Common species to use (list is not inclusive, many suited to drier sites).

Shrub species

Mediterranean saltbush	Atriplex halimus
Doryncium	Doryncium species
Bluebush	Kochia prostrata
Tagasaste palmensis	Chamaecytisus
Tree medick	Medicago arborea
Ceanothus	Ceanothus species
Mountain mahogany	Cerocarpus montanus
Shrubby germander	Teucrium fruticans
Shrubby wattles	Acacia species
Perennial lupin	Lupinus spp

Tree species

Poplars (various)	Populus species
Willows	Salix species (esp. Kinuyanagi)
Other tree crops	Black locust and various nut trees – site specific

Pasture species grasses

Wheatgrasses (Agropyron Species), Tall oatgrass (Arrhenatherum elatius), Cocksfoot (Dactylis glomerata), also fescue species and native grasses.

Pasture species legumes

Milkvetches	Astragalus species
Crownvetch	Coronilla varia
Birdsfoot trefoil	Lotus corniculatus
Sainfoin	Onobrychis viciifolia
Serradella	Ornithopus sativus
Lucerne	Medicago sativa
Sweet clover	Melilotus species
Clovers	Trifolium species

Herbs

Chicory	Cichorium intybus
Sheeps burnet	Sangisorba minor

On other sites where there is more moisture and possibly shading some of the more traditional grass-clover mixes can be used but they may require higher inputs of fertiliser to ensure sustained production.

Chapter - 8

Managed Reversion of Retired Land

8.1 LOWLANDS

Generally, the managed reversion of lowland areas take place along streambanks or wetland areas. Fencing off areas will be one of the most important factors in retirement of these areas, this is not only vital to protect plants from stock but also removes the pressure from streambanks reducing another source of erosion.

A great deal of information has already been produced about riparian management. For further details on restoration of vegetation in these areas readers are referred to the MfE publication *Managing Waterways on Farms, A guide to sustainable water and riparian management in rural New Zealand*, June 2001.

8.2 HILL COUNTRY

8.2.1 Introduction

In the early years of soil conservation, indigenous species were not favoured by catchment boards. The prevailing view was that they should not be used on farmland, either because they were too shallow-rooted to stabilise slopes, or too slow-growing to stabilise them fast, or too unprofitable a use of productive land. If a farm conservation plan provided for fencing-off standing bush remnants or reverting scrub, it was usually on a steep face or gully, viewed as having no productive value.

On the other hand, many millions of hectares of bush, scrub and tussock, reserved in the ranges and mountain lands between 1890 and 1940, were managed by the former Department of Lands or New Zealand Forest Service for the purpose of watershed protection. The view was that retaining indigenous cover in river headwaters controlled soil erosion, reduced flooding, and maintained water supply.

With growing public interest in conservation since the 1970s, the situation has changed. Many farmers now fence off remnant bush, scrub and wetland, because they perceive them as having ecological value, and as enhancing the farm landscape. It is not uncommon for landowners to fence off small areas of pasture that are difficult to farm, such as on streambanks or in gullies, and plant them up in native species.

Streambank planting of native shrubs in hill country. This site was pasture 6 years before. *Photo: D Hicks.*

Many farms already have patches of tree or shrub cover on land which has never been cleared, or which has reverted. In some instances such land is stable, though more often than not, it has been left uncleared – or has reverted – because of past instability. Retention and restoration of such areas is a low-cost soil conservation option. The key elements are:

- Construct a stock-proof perimeter fence
- Undertake regular control of animal pests
- Eradicate invasive weeds

8.2.2 Fencing

If stock are still grazed on adjacent land, a good fence is essential to success of restoration. Canopy trees may remain for many decades in the presence of browsing stock, but if seedlings are persistently browsed they do not survive; ultimately the stand will open up and collapse if regeneration is insufficient to replace the old trees.

What kind of fence to build, will depend on what a farmer can afford. A traditional retirement fence, with timber posts at 3 to 4 metre spacing and 6 to 8 wires with attached battens, is clearly the best king to ensure permanent stock exclusion. At a cost of some $10 a metre, it is also the least affordable.

Go for one of these where the landowner can afford it, or where financial assistance is available from a regional council or QEII Trust. Where not, go for a low-cost fence design – it may not be as good, but it will achieve satisfactory stock exclusion most of the time. Some low-cost options are:

- Electric fence with permanent posts at 3-4 metre spacings, and either two hot wires (for cattle) or four (for sheep). These cost about $3 a metre
- Light fence with timber posts or iron fencing standards at 5 to 10 metre spacings, with four wires (two electrified) for sheep or two wires (one electrified) for cattle; and adjustable tensioners on each wire. These cost about $1 to $1.50 a metre.

It is a good idea to have a gate in the fence; alternatively removable wooden rails between tow posts; so that if stock accidentally get inside, they can be easily removed.

Remember that a retirement fence can often be made more attractive to a farmer, by placing it in a position which facilitates grazing management. It may subdivide an existing paddock in two, giving a better rotation; or create a race that concentrates stock along a ridge or a slope-foot when mustering.

8.2.3 Weed control

Once stock are excluded from a patch of reverting scrub or bush, there is a risk that regrowth may be things that farmers don't want – blackberry, gorse or subtropical shrubby weeds – as well as natural regeneration. It is best to nip these in the bud, while infestations are still small. If they get away on a retired hill country face or in a gully, the infestations will become near-impossible to control without damaging the natural regeneration.

Knockdown herbicides such as glyphosate are effective on grasses and many woody weeds, but extreme care is needed when applying them close to regenerating natives. Where infestations are small, spot-spray using a funnel, spray wand or similar to direct spray away from seedling foliage. Residual herbicides such as oxydiazon are preferable as they give longer control before re-spraying. While more toxic, they are usually herb-specific i.e. cause minimal damage to the seedlings of woody species.

Alternatively clear rank grass and weeds by hand or slasher, to free seedling that are small and slow-growing, and leave as a mulch around their stems to help suppress regrowth.

Another option is to cut 0.5 metre diameter mats of permeable erosion control fabrics and place these around each seedling. Heavy cardboard or old carpet can serve the same purpose equally well.

A third way is to use plastic tree-guards, or off-cuts from dynex tree protectors. As well as keeping seedlings free from choking by grass, these give some shelter from wind. One dynex sleeve should provide 5 to 10 seedling-sized off-cuts. If more substantial artificial shelter is needed on windy sites, peg sheltercloth to windward on wooden stakes or iron fencing standards, but use heavy-duty material that will last several years.

8.2.4 Pest control

Protection of seedling from goats, deer, rabbits, hares and possums is essential for good natural regeneration. Shooting or poisoning will generally be needed once a year for effective pest control. Additional protection may be obtained by applying repellent chemicals e.g. egg-paste to seedling. Doing this is low-cost but laborious, so is an option only in small retirements.

Fenced-off retirements at least have the advantage that pests in them can be easily targeted. Depending on the type of vegetation and when different animals feed on it, it may be possible to hit them hard once a year, instead of having to spread effort. Some regional councils have charts showing what times of year possums feed on different plants, and similar information may be available for the other animal pests.

8.3 REVEGETATION

Where gullies or steep faces are too unstable to contemplate spaced planting of poles in pasture or close afforestation with commercial timber species, the options are either to fence, and await natural reversion; or to assist it by planting native seedlings in the rank grass. The first option is clearly preferable because it is low-cost. In wet regions like Northland, Auckland, the Waikato, Taranaki, Nelson, Westland and Southland closed-canopy scrub will cover the site within 10 to 20 years. However it will take between 20 and 30 years for a closed scrub canopy to form in seasonally dry regions (Gisborne, Hawkes Bay, Wairarapa, eastern coastline of the South Island); and longer than 30 in dry inland regions of the South Island. Where faster revegetation is desired, the second option becomes necessary.

Key elements of successful indigenous revegetation are:

- Fencing
- Choice of species
- Planting
- Post-planting maintenance
- Pest control
- Weed control

For three of these – fencing, pest control and weed control – details are the same as for restoration of bush remnants and reverting scrub. Readers are referred to the preceding discussion of these topics. The remaining three are discussed below.

8.3.1 Choice of species

What species to plant, may be decided as much by a landowner's personal preference or by ecological considerations i.e. what the original vegetation was, as by species' effectiveness for soil conservation. The QEII Trust's Revegetation Manual (Evans 1983) and DOC's Trees for Tomorrow (Simpson 1995) are standard references for ecological restoration. The Plant Materials Handbook, Volume 3 (Van Kraayenoord and Hathaway 1986) is still a standard reference about use of native species for soil conservation. However in the years since its publication, more has become known about their suitability for this purposed in different environments. The following list gives species that various regional councils recommend for planting on hill country slopes (including gullies).

Tree planting of gully system and retirement from grazing. Photo: Wellington RC.

Note that it is a nationwide list. Some of the species on it are suited to a mild climate so will establish well in northern parts of the country but may fail in cold, frosty conditions farther south. Equally, some of the hardy southern species may not thrive in hot, dry northern summers. Use local knowledge – observation of what grows well naturally in a region – when selecting plants from the list. Be open to other species as well – there may be ones that, though restricted in their natural distribution, are suited to soil conservation at particular localities.

Refer to Table 8.1 on the following page.

8.3.2 Planting

Obtain seedlings from nursery stock, preferably propagated from local seed sources. These have a better chance of survival, as they have been selected from trees that are adapted to local soils climate.

Bare-rooted seedling are easiest to handle, but have to be planted soon after supply. Potted seedlings can be held for longer, but are bulky. Root-trainer seedlings can be held for a long time, and are light to shift, but do not have much foliage which lessens their chances of survival (though good results can often be obtained).

Seedlings should preferably be not shorter than 20cm and no taller than 50cm, as smaller or larger seedlings can be difficult to establish. All plants should be hardened off in the open for about one month prior to planting.

Table 8.1 Species for Revegetating Hill Country Slopes.

		Tolerance to soil	**Wind**	**Drought**	**Frost**	**Damp**	**Value for slope stabilisation**
Agathis australis	kauri	m	m	m	l	m	*
Alectryon excelsus	titoki	m	m	m	m	m	*
Aristotelia serrata	makomako	l	m	l	h	m	?
Beilschmiedia taraire	taraire	l	m	m	l	m	*
Beilschmiedia tawa	tawa	l	m	m	m	m	*
Brachyglottis repanda	rangiora	m	m	m	m	l	?
Carmichaelia spp	broom	l	l-h	m-h	h	l	@
Carpodetus serratus	putaputaweta	l	m	m	m	l	?
Cassinia spp	tauhinu	m	m	m-h	m-h	l-h	@
Coprosma lucida	karamu	m	m	m	m	h	?
Coprosma parvifolia	leafy coprosma	l	m	m	h	l	?
Coprosma propinqua	mingimingi	l	m	m	h	m	?
Coprosma repens	taupata	h	h	m	m	l	?
Coprosma robusta	karamu	m	m	m	m	m	?
Cordyline australis	cabbage tree	m	h	m	h	h	@
Coriaria spp	tutu	l	m	l-m	m	l	@
Corokia cotoneaster	korokio	m	h	m	h	l	?
Cortaderia spp.	toetoe	m	h	h	h	m	@
Corynocarpus lavig.	karaka	h	h	m	l	l	*
Cyathea spp.	tree fern	l	l	l	h	h	@
Dacrydium cupressinum	rimu	m	m	l	m	m	*
Dacrycarpus dacryoid.	kahikatea	l	m	l	m	h	*
Dodonea viscosa	akeake	h	h	h	m	l	@
Dicksonia spp.	tree fern	l	l	l	h	h	@
Entelea arborescens	whau	m	m	m	l	m	@
Fuschia excorticata	tree fuchsia	l	m	l	m	m	@
Griselinia littoralis	broadleaf	h	h	m	m	m	*
Hebe spp	hebe	l-m	m-h	l-m	m-h	l-m	?

Hoheria spp	lacebark	l	m	l-m	m	m	@
Knightia excelsa	rewarewa	l	m	m	m	l	*
Kunzia ericoides	kanuka	l	m	h	h	m	@
Leptospermum scopar.	manuka	m	m	h	h	h	@
Macropiper excelsa	kawakawa	l	l	m	l	m	?
Melicytus ramiflorus	mahoe	l	m	m	m	m	@
Metrosideros spp	rata	l-h	m-h	m-l	m	l	*
Myoporum laetum	ngaio	h	h	m	m	l	@
Nothofagus fusca	red beech	l	m	m	h	m	*
Nothofagus menziesi	silver beech	l	m	m	h	m	*
Nothofagus solandri	black beech	l	h	h	h	m	*
Nothofagus cliffort.	mountain beech	l	m	h	h	m	*
Olearia spp	leatherwood	h	m	m	h	l	@
Paratrophis banksii	towai	l	m	l	l	m	*
Phormium tenax	flax	h	h	m	m	h	@
Phyllocladus glaucus	tanekaha	m	m	m	l	m	*
Pittosporum colensoi	black mapou	l	m	m	h	m	@
Pittosporum eugenioid.	lemonwood	l	m	l	m	m	@
Pittosporum crassifol.	karo	h	h	m	m	l	@
Pittosporum ralphii	karo	h	h	m	m	l	@
Pittosporum tenuifol.	kohuhu	l	m	m	h	l	@
Plagianthus spp	ribbonwood	l-h	m-h	m	m	m-h	?
Podocarpus totara	totara	l	m	m	h	m	*
Podocarpus halli	mountain totara	l	m	h	h	m	*
Prumnopitys ferruginea	miro	l	m	m	h	m	*
Prumnopitys taxifolia	matai	l	m	h	h	m	*
Pseudopanax arboreus	five-finger	l	m	m	m	l	@
Pseudopanax lessoni	five-finger	l	m	m	m	l	@
Pseudowintera axill.	horopito	m	h	m	m	m	@
Schefflera digitata	pate	l	l	l	m	m	?
Solanum aviculare	poroporo	l	m	h	l	m	?
Sophora spp	kowhai	l	m	m	h	m	@
Weinmannia racemosa	kamahi	l	m	m	m	m	@
Vitex lucens	puriri	m	m	m	l	m	*

Abbreviations: l = low tolerance; m = moderate tolerance; h = high tolerance; @ = fast growth and root spread; good value for slope stabilisation; * = slower growth and/or root spread; lower value for slope stabilisation; ? = limited growth and/or root spread; some value for slope stabilisation

In northern regions, seedlings are best planted in May or June, so that they establish over the winter and spring months before summer drought sets in. In central and southern regions, seedlings are best planted in mid-winter or early spring, when the plants are dormant and after they have been hardened off by frost. Any species vulnerable to out-of-season frosts should be planted as late as possible, in September or October.

Grass and weeds should be sprayed or hand-weeded 2 to 4 weeks prior to planting. A 0.5 to 1m diameter spot at each planting site is usually sufficient. This gives seedlings a chance to put on a spurt of growth before the grass and weeds grow back. Where there is concern about spray side-effects or residues, use amine-based salt preparations such as glyphosate – they have low toxicity and break down fast.

Dig a planting hole that's large enough to accommodate the seedling's roots with plenty of loose earth around and underneath once back-filled. Take care that the seedling isn't too deeply buried. Seedlings should be planted with potting mix, but ensure that the potting mix is not exposed. Keep it about 2cm below ground level and cover with soil or mulch.

Direct seeding may be worth trying, as an alternative to plantation of seedlings. Australian landcare groups report good results with this technique. In New Zealand there has been little experience with species other than manuka and kanuka, which may be established by scattering seed-pods or laying brush with ripe seed-pods still attached.

Establishment can be helped by fertilising with one or two 25g magamp pellets in the planting hole. Alternatively use 25-50 grams of granulated magamp. Magamp is preferable to urea or nitrolime which can easily burn plant roots. It is better to under-fertilise than over-do it; too much fertiliser around its roots will kill a plant.

For indigenous seedling, an initial spacing of 2.5 to 3.5 metres is recommended, so that growing trees establish reasonably dense ground cover.

Most native species are naturally self-thinning, but if timber production is contemplated, hand-thinning to 5-7 metres at 20-plus years may help improve growth rates. Fast canopy closure, with suppression of competing weeds or grass within 2 to 3 years, can be achieved by planting at 1 metre spacings. However this entails 10,000 seedlings a hectare, so is feasible only on small restoration sites (or projects with big budgets e.g. road earthworks).

8.3.3 Post-planting maintenance

Post-planting maintenance can be crucial for seedling survival, but may not be practical due to other demands on time. The options are:

- Only plant what can be cared for, or
- Plant at a higher density to allow for partial planting failure, and let nature take its course

It is unrealistic to expect 100% success with any tree planting, but post-planting care and attention can greatly improve their chances. When good-quality seedlings are planted and protected, 7080% usually survive. Where animal pests are present, they are the single largest causes of tree mortality, and much higher percentages may need to be replanted.

In inland regions, blank (re-plant) gaps in August or September, because May to July may have been too early for frost-tender seedlings; or too dry on free-draining soils. In coastal districts, blank in May or June, so that seedlings have a chance to establish before drought sets in the following summer.

If planting indigenous timber trees with a view to harvest, it is a good idea to retain some documentary evidence that they are planted and not natural regrowth, so that harvest isn't precluded by existing or future forestry legislation.

If planting indigenous trees in the hope of a commercial return, pruning is advisable to improve growth form, and avoid knotty sawlogs. Most New Zealand publications on silviculture deal specifically with radiata pine, but the techniques they recommend could be applied equally well to many timber species, including natives. For high-quality clearwood, some farm foresters make a practice of annually pruning small side-growth branches on their native timber stands.

Where trees are planted close to a valley-bottom stream in hill country, they need to be clearly separated from the stream, so they they can be felled without any risk of the channel being disturbed by logging machinery, or blocked by slash. Even where there is no intent to harvest, this is a good idea, because the channel still needs to be managed so as to maintain its flood capacity. Consider planting patterns that restore the natural flow path of floodwater. Trees that fall in may block the flow path, and need to be removed. Dense plantings, or weed regrowth enclosed by plantings e.g. self-sown willows, blackberry or gorse, may also block or deflect the flow path. Any debris which has lodged in the bed may need to be removed, if it is likely to impede passage of floodwater or direct it against banks which have been planted.

8.3.4 Nurse crops

Given that many indigenous plants are slow-growing, a nurse crop of exotic shrubs may be worth considering, to provide quick ground cover and shelter. The plants on this list form an open canopy, which admits enough light for native seedlings to grow underneath, and eventually dies out when it is overtopped by the emerging natives.

Albizzia spp.	Brush wattle
Banksia spp.	Banksias
Callistemon spp.	Bottlebrush
Chamaecytisus palmensis	tree Lucerne
Lupinus arboreus	tree lupin

Tamarix chinensis tamarisk

Tree Lucerne in particular is good fodder for native birds, and can enable natural revegetation from seeds they bring in. Shrubby acacias are excluded from the list because many species are self-seeding and can from dense thickets. Casuarinas are excluded because there is some concern about their weed potential, though one species, river she-oak (Casuarina cunninghamii) may be suitable.

8.3.5 Concluding Remarks

That indigenous plant materials can be effective slope stabilisers in hill country, has been demonstrated by numerous storm damage surveys carried out from the 1960s onwards. Closed-canopy scrub and bush – whatever its species composition – reduces mass movement and gully erosion by 90% or more, compared with adjacent land in pasture (Clough and Hicks 1993, Hicks 1995). Reverting scrub, with an as-yet open canopy and discontinuous root mass, achieves reductions of up to 50%.

It is true that there is limited scope for spaced planting of indigenous species in grazed hill country pasture, unless the seedlings are protected or the pasture lightly grazed. Close planting of indigenous timber trees on unstable hill country, while occasionally practices, remains a less attractive commercial option than fast-growing exotics. It is the third category of hill country – unstable sites with low potential for grazing or forestry – where indigenous revegetation will be increasingly used as a soil conservation measure.

8.4 MOUNTAIN LANDS

This is the practice of fencing out an area of severely eroded high country (usually LUC VII and VIII), which is often at the top of a mountain range. In some circumstances the topography of the area may enable stock to be completely removed without the need for large lengths of retirement fences

Furthermore, to assist in the revegetation process of the retired land all stock that were removed will require alternative grazing elsewhere i.e. offsite feed. This is usually provided from the lower altitude, more productive land i.e. which may be AOS & TD or irrigated lucerne to carry the same or equivalent displaced stock on.

Retirement schemes were not really advanced at all until the Soil Conservation & Rivers Control Council provided grant incentives to farmers to carry out the practice from about 1955 onwards. This ceased in 1992 when grant monies for soil conservation purposes were terminated by government.

Today the works still continues on private land but are at a slower rate unless it is part of the Crown Land Tenure Review process on (Pastoral Leasehold land) where the high altitude or severely eroded lands are retired and transferred to the Crowns estate to become the responsibility of the Department of Conservation.

The same land use change can occur on Freehold land where severely eroding land exists and needs to be destocked to assist in the natural or man assisted revegetation programme. There may be opportunities to involve the Queen Elizabeth II Trust if there are unique areas of tussock grassland and bush within the same block.

The most important factor in any retirement block is to ensure that ongoing animal and plant pest control is carried out as well as regular fence maintenance. It is vital that proper fire management control such as firebreak tracks and watering facilities are in place in case any escaped or wild fires occur which will damage the vegetation recovery programme.

Runoff Control & Dewatering

Chapter

Chapter - 9

Runoff Control Practices – Lowland

9.1 INTRODUCTION

The practices described apply to the rural environment and are particularly applicable on arable lands. Some practices can also be used to control siltation and erosion in urban development.

Some of these practices have not been used much in New Zealand (e.g. terracing), but are suited to intensive land use on slopes of 12° and less. The earthwork control practices should be constructed in the early phases of erosion control, e.g. graded furrows, banks as they become permanent features. The other soil conservation practices can be integrated with them afterwards, i.e. conservation tillage, cultivation, soil/pasture aeration and others.

Many of the runoff control practices have been used in New Zealand from the 1950s. They were first tried on soil conservation reserves and trial sites as The Pisa Flats (Central Otago), the Adair Reserve (South Canterbury), Glenmark Catchment (North Canterbury) and the Earnscleugh Reserve site (near Alexandra). The techniques were generally successful on low terraces and dry downland sites. From the 1970s onwards these practices seemed to loose favour as farm development and grant monies were directed into pasture improvement.

9.1.1 Contour furrows

These can be simple systems constructed on a paddock either on the contour or with a slight gradient (may also be called a diversion or drainage furrow).

Contour furrows are used to detain runoff on hill slopes. They break slopes up into a number of short lengths, preventing runoff reaching velocities that could cause rilling or sheetwash. Dung, fertiliser and soil are also retained in the furrows, creating fertile, moist conditions. More useful in drier environments where mass movement erosion is not a problem.

Graded or diversion furrows are used to divert water away from problem areas, e.g., slips, gully heads, gateways, or slopes where sheet/rill erosion occurs taking the water to safe disposal sites. They are useful to irrigate sunny knobs or ridges. Such sites are also ideal outfalls, since they tend to spread water out, rather than concentrating it in one place as occurs in a gully. Graded furrows can be used to increase the effective catchment area of a pond or dam.

Graded furrows for Drainage can be an effective and inexpensive method of draining slopes over 10°, where other methods are impractical. Put in at 10 m intervals, they may lead to a safe disposal site or grassed waterway.

Graded Furrow, North Otago. ***Photo: Otago Regional Council.***

9.1.2 Factors to Consider in Design

Slope

More water will be detained on flatter land than steeper land. Furrows on steeper slopes must be installed closer together because they have less water-holding capacity. Furrows can be installed effectively on slopes of up to 25%.

Climate

In areas where high intensity rainfall frequently occurs, furrows would be spread closer together.

Soil Type

Fewer contour furrows are required on free-draining soils where infiltration is rapid. In clayey soils or wet conditions, furrows should be graded to a safe disposal site.

Size of furrow

A furrow made by a grader may detain 23 times the amount of water detained by a furrow constructed with a single-furrow plough.

Vegetation

Crops and lucerne which may have a high percentage of bare ground and rapid runoff, require furrows to be closer together than will grassed pastures.

Access and Cultivation

Contour spacing should allow for vehicular access routes if these have been the custom – a compromise may sometimes be necessary between desirable contour spacing and actual practice. Contours should also be spaced and positioned for future cultivation.

Machinery

Furrows can be formed by a number of different machines, and their capacity is strongly influenced by the resulting cross-sectional area. When a two furrow plough is used to form the pasture furrow, they can be ploughed so that the second turf is placed on top of the first. Construction by grader type machines is the preferred method on slopes less than 15% and plough furrows on slopes over 15%. For steeper slopes, bulldozers have been used, but it is more difficult to achieve the critical grades required.

9.1.3 Practical Points and General Recommendations for Constructing Furrows

Contour furrows

- A system of contour furrows can be expected to detain 25mm of runoff.
- Pasture furrows are useful on slopes of 7° to 18°. On flatter arable areas, broad-based terraces that are no impediment to machinery are more useful.
- Generally, vertical interval between furrows is 2 – 4m, the steepest slope point. A 2m vertical interval is equivalent to 7-10m horizontal interval on a 10°-12° slope. On 25% slopes, the interval would be 8m
- Furrows should be compacted by running the rear tractor wheel along them when returning empty to the start of the next contour line.
- Furrows should be seeded and fertilised once constructed so they quickly get a protective cover of vegetation. In lucerne, furrows are best sown in sub. clover.
- Cattle should be kept off newly furrowed paddocks furrows. May not be suitable on deer farms as deer like to wallow and damage areas of some paddocks.
- Furrows should be broken every 50m to 100m in a zig-zag fashion down the contoured slope. This prevents overtopping and scouring occurring at any one place.

- Do not construct contour furrows on land prone to slips or tunnel gully erosion, as increased infiltration can worsen the problem. Graded furrows can be constructed only above or below these areas.
- Maintenance with a shovel over the first year is required.
- Furrows become filled up with soil, dung and vegetation after 3-4 years. The decision to reconstruct these or cut new furrows depends on whether the mitigation of soil loss and runoff has been achieved.

Graded Furrows

- Normally a grade of 1:80 should not be exceeded or scouring of furrows could result.
- A quickset level or some other accurate surveying equipment should be used for graded banks and similar structures since fall may be only 1:200 or 1:400.
- Graded furrows can be spilled out on dry knobs or ridges. These are ideal disposal sites since water is spread rather than concentrated as it is in a gully. Soil moisture levels are also improved on these normally dry sites. Regular maintenance is vital.

9.2 BROAD BASE TERRACES AND ABSORPTION TERRACES

These structures are used on flatter cropping land (2° – 7°). These are commonly 9 to 12 metres wide and 300mm high and is no impediment to the passage of machinery.

On gentle slopes (2° – 5°), a small mound will detain quite a large amount of water. However, with increasing slope, the amount detained falls off quickly.

Like pasture furrows these structures break long slopes into short lengths, encouraging absorption and preventing rilling.

On slopes steeper than 5°, terraces over 100m should be graded at 1:300. Excess water is slowly led off the land without scouring to a grassed waterway. Storage behind terraces on these slopes is insufficient to cope with heavy rain. If they are not graded overtopping may occur. (Since grades are slight, it is essential to use a Quickset level, Abney level or other surveying equipment).

Machinery for Broad Base Terraces

Terraces are built up with a grader and/or a whirlwind terracer, which is a tractor drawn implement which cuts a furrow and feeds it into an auger revolving at variable speed. By changing the auger's speed, soil can be thrown a distance of up to 8m.

Limitations

Terraces are not used extensively in New Zealand, in contrast with densely populated places overseas where flat land is scarce. Orchards and vineyards have been successfully established on terraces constructed along steep slopes (20°) in the Heathcote Valley, eliminating sheet erosion under the trees, also for kiwifruit orchards in the Tauranga district.

Terracing is very expensive and requires careful stockpiling of topsoil. Surveying is vital to ensure correct gradients and safe outfalls are established. Do not use this technique where slip erosion can occur. On slopes that may be susceptible to water logging from ponded water, graded terraces are more appropriate.

Graded banks on moderately steep Bentonitic mudstone country. ***Photo: Gisborne DC.***

9.3 GRADED BANKS

9.3.1 Description/Purpose

Graded banks are a series of earth banks/channels formed on long slopes at very slight grades, to control the surface stormwater flow and divert the runoff to stable grassed waterways. The purpose of the graded banks is to dispose of surplus overland flow so that the risk of rill or gully initiation is lessened. They also improve drainage on clay soils.

Application

Graded banks are generally smaller than diversion banks and broad-based terraces. The banks are normally constructed using a grader on slopes of 12-15°.

Graded banks can be constructed on steeper slopes using a bulldozer, but the grades are more difficult to accurately control. Over the years, a range of machinery including ploughes and whirlwind terracers have been used to install graded banks.

Ideally, the grades on the banks should be as low as possible, depending on how far the water needs to be diverted. Typically, grades are between 1:80 to 1:100. However, for short distances, grades can be as steep as 1:60 as long as scour velocities are not reached. Graded banks are installed at spacings as close as 20 m apart. They need to be accurately surveyed and grassed as soon as possible after construction. The outlet for the graded banks needs to be a waterway specifically planted (grass/ trees) and managed so that it can handle the volumes of runoff without eroding. Controlled waterways should ideally spread water so that it is not concentrated.

Graded banks have been successfully used in Northland to control gullying on gumland soils, and on in the Wither hills (Marlborough) for tunnel gully control by diverting surface runoff back to ridges to spread the water flow. They have also been used in the Gisborne District in conjunction with pole planting and dewatering to control earthflow and gully erosion. In this case they were constructed using bulldozers, and poles were planted on the outside of the graded bank to help stabilise the works in the medium to long term.

9.4 INTERCEPTION DRAINS

These perform the same function as diversion channels/bunds (see Runoff Control on Earthworks Section 16) for greater design detail if required.

Description/Purpose

Interception drains are site specific measures to intercept and convey runoff at non-erosive velocities away from cropping soil to a stable outlet. They are permanent drains, and should be part of an integrated system, as opposed to contour drains which are temporary.

Installation

Grade should be less than 2 % otherwise the channel will erode and need to be stabilised with rock or erosion control fabric. The outfall may need to be protected against erosion with rock, or a flume. Because of their permanent nature and the small margin for error, these channels must be surveyed to obtain an even grade. An abney level, inclinometer or similar should be appropriate for this purpose.

Earth bunds can be used to control and direct runoff. They need to be well compacted. Any excavated channel or bund wall should be grassed to prevent scouring.

No criteria are presently advocated for a diversion channel in a cropped field. If one is required, then it is suggested that design be for the 1 in 20 year (5% AEP) storm.

Limitations

Because of the interception drain's low grade and the well-aggregated soil type, sediment will often deposit. Sedimentation can quickly compromise the capacity of the drain to transport flow and cause localised damage to crops at break out points.

Sandbags used to control drain velocity. *Photo: Franklin Sustainability Project.*

Care needs to be taken in their placement so that the drains can be accessed and any sediment removed. Constructing the channel on a steeper grade so that sediment is not deposited will result in the channel itself eroding and requiring erosion control. Check dams made from rock (see Earthworks Chapter 16), sand bags etc are often necessary to control channel velocity. Sandbags filled with a mixture of sand, gravel and concrete are very effective in steep sided drains. The concrete will set and hold the sandbags together (use UV resistant bags if possible). Check dams constructed in this situation will also act as sediment retention measures because velocity reduction results in sediment deposition on.

Maintenance

Interception drains need to be inspected regularly, particularly during and immediately after heavy rain, to ensure that the drain is working correctly and is not blocked. Deposited sediment needs to be removed and any necessary repairs to the drain should be undertaken. The outfall should be checked to make sure that is free from erosion. Check that machinery movement through the site hasn't damaged the channel.

9.5 HEADLANDS

These can be significant sources of runoff due to severe soil compaction.

Description/Purpose

Headlands are used for access and machinery turning at the ends of cropped fields. They can be designed to intercept and divert runoff to permanent interception drains. They are still used in the normal manner for machinery access.

Installation

They should be constructed at no more than a 2% grade (to avoid scour). They can be sloped back to form a benched headland. Alternatively a wide shallow "V" channel can be formed between the headland and end of the rows (this latter type requires more shaping). They can also be formed within paddocks, to break up the length of long paddock runs. If constructed to a broad shallow design, machinery can be easily driven across. Ideally the headland should be grassed (and it can then be on a slightly steeper slope as a grassed surface is more resistant to erosion).

Limitations

Common problems are insufficient capacity, erosion prone outfalls, too steep (so they erode), or too shallow a grade (so they don't flow or deposition compromises their capacity). They have often been constructed so they discharge straight to an accessway and then to a road or road side water table.

They may not have been installed where needed through paddocks, because the resultant uneven ground can affect machinery operations.

Maintenance

Remove any accumulated sediment. Re-grass as required. Repair as necessary.

9.6 CONTOUR DRAINS

These perform the same function as those described in Section 15.3, Runoff Control on Earthworks.

Description/Purpose

They are temporary drains used to intercept runoff across slope in a paddock and channel it to permanent interception drains.

Installation

Contour drains should be as short in length as possible and constructed at no more than a 2% grade (to avoid scour of the channel). As a rule of thumb they should be about 300 mm deep and from 20 to 80 metres apart. Drain spacing should be 20 metres apart for slopes of 10% or more, and can increase to 30 and then 50 metre spacing as slope gradient changes to 3-10% and to less than 3% respectively. As a general rule, contour drains should never be more than 80 metres apart. The actual positioning should be determined by the presence (or otherwise) of an erosion proof outfall. They should be put in immediately after sowing a crop.

Contour drains. *Photo: Franklin Sustainability Project.*

Limitations

Care is needed with contour drains because they concentrate site runoff. Insufficient capacity, erosion prone outfalls, and construction on too steep a grade (so they erode) or too shallow a grade (so they don't flow) are common problems. They are often too far apart, or do not discharge into a permanent drain. They are often not installed by growers where needed, because the uneven ground hinders machinery movement and spraying operations.

Maintenance

Contour drains should be inspected during periods of prolonged rainfall and as soon as possible after heavy rainfall. Any necessary repairs should be made.

9.7 RAISED ACCESSWAYS

Description/Purpose

Raised accessways are constructed from roads into fields, to facilitate machinery and truck entry. A raised accessway is not a runoff conveyance system, it is a measure by which runoff is diverted away from this heavily trafficked and vulnerable spot.

Installation

Accessways should be placed away from the lowest point in the field where water naturally flows. They should be raised above the surrounding area so runoff is diverted sideways into the nearest interception drain. This can be done by placing aggregate on the accessway to mound it up, this gives all weather access and reduces the potential for future lowering of the accessway. Any accessway that goes directly onto a road will need to be piped to allow roadside drainage. The size of culvert should be checked with the local council or drainage engineer.

Figure 2(10) raised accessway Photo – Photo: *Environment BOP.*

Maintenance

Maintain the raised nature of the accessway to ensure runoff cannot flow directly to the road. Keep the culvert under the accessway open and clear.

Silt traps. ***Photo: Franklin Sustainability Project.***

9.8 SILT TRAPS

Description/Purpose

Silt traps are pits dug at intervals along interception drains. Their purpose is to retain sediment on-site and minimise off-site sedimentation.

Installation

They should be used in combination with measures that reduce runoff. The smaller the quantity of runoff directed through these measures, the less sediment they need to trap. Silt traps can be constructed either by excavating pits or constructing bunds, and passing runoff through them prior to discharge off site. Excavations can consist merely of a hole, preferably longer than it is wide and with the outlet at the opposite end to the inlet. Earth bunds can be used as well, or in conjunction with excavations, to form a silt trap.

There are currently no guidelines on trap sizing. On strongly aggregated soil types, it is probably appropriate to excavate a series of scoops along a drain rather than construct a more formal silt trap at the lower end of a paddock. However, for less well aggregated soil types, and assuming that there is an offsite need for sediment to be retained, then sediment retention ponds should be used. A drainage pipe, such as those detailed in earth bunds (see Earthworks Chapter) can be used to drain the pond or depression, or the pond should be kept shallow enough so that it is not a safety hazard. Silt traps should be located in side drains as well as permanent drains. They need to be located away from turning areas for safety reasons.

In permanent drains, small check dams constructed out of rock, timber or erosion control fabric can be used to retain sediment as well as to reduce the velocity of water in the drain. Scoops (slightly deeper excavations) at 15 – 20 metre intervals along a drain can also be used.

Limitations

Silt traps quickly fill with sediment and will need regular digging out. Minimising the quantity of runoff will reduce the frequency at which this needs to be done.

At present there is no specific design requirement for silt traps – they are merely "holes on the ground". Room is always at a premium on grower's properties and these measures are normally "squeezed in". They may, therefore, quickly fill with sediment and lose their effectiveness during a large storm. Some recognition of catchment size, inflow velocity and silt trap capacity, undertaken at initial design, would improve their effectiveness.

The reason that they work as well as they do is because of the very strongly aggregated nature of the granular loams. This allows the particles to settle easily when runoff velocity diminishes. Silt traps will not be as effective on other soil types that readily break down to clay and silt sized particles, because their typically small capacity will not allow sufficient settling time to retain the fine particles. Sediment retention ponds (see Earthworks Chapter) would be more effective in situations where the soil is easily broken down to clay and silt.

Maintenance

Silt traps should be regularly cleaned of sediment so their effectiveness is retained. Any drainage pipe should be inspected to ensure that it is not blocked and is functional.

9.9 SUBSOILING – AMONGST TREES

Subsoiling involves the ripping of soil to a depth of about 460 to 600mm to shatter the compacted pan, particularly under tractor wheel marks, as the compacted soil obstructs water infiltration.

Application

Subsoiling can be undertaken using a D4 bulldozer with trailing twin rippers hydraulically mounted, or mounted on a hydraulic toolbar. Soil conditions need to be dry enough to shatter, but not so dry that the machine cannot rip to the required depth. Work is normally undertaken in spring/early summer (October/November) or early winter (May/June). Subsoiling runs could be up to 100 m long on slopes of less than 8°, but limited to 40 m on steeper slopes. Two rips per row are necessary to break the compacted layer under tractor tyre marks.

Subsoiling was not wholly accepted by landowners as being necessary, and their concerns related to the questionable drainage benefits as well as the possibility of killing roots of producing trees. Although the Nelson Catchment Board recommended that more investigation be undertaken to quantify the effects of subsoiling, staff had little doubt that the practice was absolutely necessary to improve the infiltration and provide a basis to initiate underground drainage patterns.

9.10 TILED DRAINAGE BENEATH WATERWAYS

These include a 100 mm field tile (drainage pipe) in a trench (170 mm wide and at least 900 mm deep), and backfilled with coarse drainage metal, then grassed. The tiled grassed waterways collect water from the subsoiling and drain water via the drainage metal and field tiles safely down to the outlets. The system also provides drainage for the wetter soils of the flats and valley floors.

Installation of tile drains through orchard. **Photo: Nelson Catchment Board.**

Main and outlet waterways usually follow natural flow paths and need little shaping to function effectively. The depth of drainage metal provides a greater capacity for drainage, and creates a break in the hard pan, allowing better infiltration to occur across the orchard. Water is able to rise as the volumes increase, and some surface flow may result during peak flow conditions. This surface water infiltrates into the drainage metal and tile drain as the flow decreases. Approximately 3.5 to 4 cubic metres of drainage metal is required for every 20 metres of trench.

Completed grassed waterway. Photo: Nelson Catchment Board.

9.10.1 Grass waterways beneath trees and along waterways

This practice includes the grassing of bare areas, particularly following subsoiling and tile draining, to stabilise the ground surface and reduce the risk of rilling and gullying.

Grassing down is normally carried out in the winter following subsoiling. The grassed waterways need to be maintained through to the outlets. This may create management problems if there is cultivation carried out on flatter land prior to the grassed waterway ending. All drainage systems need to be graded and vegetated so that they are able to convey stormwater runoff flows without scouring.

Grade line system for new orchard on rolling land. ***Photo: Nelson Catchment Board.***

9.11 GRADE LINE (CONTOUR) PLANTING

The grade line system involves planting trees along the contour on sloping land in parallel and on a similar grade to the subsoiling and tile drains. The drainage system feeds into tiled grassed waterways and then to natural channels.

Early attempts to plant orchards on the contour failed because the wide fluctuations in spacing prevented efficient spraying. Subsoil drainage allows the spacing between trees to be more uniform.

9.12 HEDGES

Hedges can act as natural barriers to retain sediment. They can be effective at the bottom of a sloping paddock, between a paddock and drain, and when combined with a headland, can be used to intercept and divert runoff from upslope. They are a permanent measure and are cheap to establish and maintain.

Hedges should be trimmed annually to encourage thick growth, particularly at ground level. Any gaps should be blocked or filled with young plants.

9.13 CONTOUR BANKS

Description/Purpose

Contour mounds or banks are similar to contour furrows except they are spaced further apart and are twice the size.

Application

Contour mounds can be used in situations where the close pattern of contour furrows creates management problems such as if the area is to be cultivated frequently. Contour mounds can be constructed on similar slopes to pasture furrows, and while not as efficient as the pasture furrows, they still provide for good moisture retention, and runoff control.

9.14 BROADBASED TERRACES

Description/Purpose

These are structures designed to detain surface runoff on flatter sites (3-12% slopes), and safely convey the runoff to a safe disposal point, normally a grassed waterway.

Application

Broadbased terraces have a large storage capacity, and usually have such a shallow batter slope that machinery and implements are able to traverse them with ease. They are normally used on relatively flat country that is intensively farmed. Broadbased terraces can be constructed with a range of earthmoving machinery. In New Zealand, a special-built machine, the "whirlwind terracer", was trialled in the South Island at Invermay Research Station, and successfully used in the Otago region during the 1950s and 1960s.

Chpater - 10

Dewatering Techniques for Deep-seated Mass Movements

10.1 INTRODUCTION

The feasibility of stabilising mass movements varies according to the form of mass movement involved. In general, **Earthflows** occur in the weathered regolith and so are usually between 1 and 3 metres in depth. However, in deeply weathered crush zones they can be much deeper. The weathered mudstone in which they form frequently contains a high proportion of the swelling clay, montmorillonite (bentonite).

Slumps, with their semi-circular failure plane, form in deep relatively uniform cohesive materials, including unconsolidated deposits such as clays and weaker tertiary sediments. They may be ten's of metres deep at their greatest depth.

The depth of a **Slide** will frequently be determined by the depth of a particular sedimentary layer that may have lower shear strength or different hydraulic properties.

Various forms of mass movement may occur in the same failure. For example the lower part of a slide may be disrupted during failure to form an earthflow or debris flow, while the upper section may contain the rotational features of a slump.

From these definitions it can be seen that a deep earthflow is unlikely to be as large a feature as a deep slump or slide and there is therefore a greater possibility of improving the stability of an earthflow.

Deep slumps and slides are major features that are challenging even with the unrestricted use of civil engineering techniques. This is illustrated by the expense of the stabilisation works in the Cromwell Gorge. While this approach lies outside the scope of this handbook, the geotechnical principles involved apply equally well to smaller mass movements.

10.1.1 Geotechnical Principles of Slope Stabilisation

The stability of any slope is a balance between restraining forces (the shear strength of the soil) and destabilising forces (gravity). Both of these forces are affected by the angle of the slope so that the opposing forces in a gentle slope are less than those in a steep slope. Where tectonic uplift is still occurring, and streams are actively down-cutting, the slope angle will be determined by the strength of the material it is composed of. Weak material will only support a gentle slope (after its toe has been removed by stream erosion) while very strong rock will form vertical bluffs.

In effect this means that many slopes where there are high rates of uplift, including North Island East Coast tertiary sediments, are quite close to failure in their natural state. Only minor changes in the strength of the soil or to shape of the slope (e.g. the removal of the toe of the slope by stream erosion or road cutting construction) may cause failure to occur.

It is convenient to regard soil shear strength as having two components:

Cohesion is not affected by the weight of the soil overlying it and also remains relatively constant over short periods of time. It can be increased by the presence of the roots of plants and can also be reduced over time by weathering processes. In the case of weakly cemented mudstones containing swelling clays the most rapid weathering process is excessive drying. Shrinkage forces break the existing bonds and re-wetting allows swelling to occur more freely than before. In one example near Gisborne a freshly exposed hard mudstone slope became a moving earthflow in only 11 years. Vegetative cover and shade is therefore very important to preserve cohesion.

In another well documented situation, railway cuttings in Britain failed up to 80 years after their construction due to progressive failure of the slowly weakening clay material.

The **frictional** component of shear strength is dependent on the **normal** force (at right angles to the shearing force) and so greater frictional strength is mobilised deeper in the slope. Opposing this normal stress is the pressure of water in the pores of the soil (the **pore water pressure**). Consequently an increase in the pore water pressure reduces the normal force which in turn reduces the shear strength of the soil.

10.1.2 De-watering

The aim of de-watering is to manipulate pore water pressure as an effective way to increase the shear strength of the soil. The essential principle is to minimise the volume of water getting into the slope and maximise the volume leaving the slope. The means of doing this will vary with the type of mass movement being stabilised but surface drainage, subsurface drainage, surface reshaping, and afforestation are basic tools.

Where agricultural or forestry land is at risk and some land movement can be tolerated the lower cost approaches described here can be adopted. Where assets such as buildings and roads are affected by the movement a more comprehensive engineering approach would be warranted for the higher level of protection it might provide.

10.1.3 Overview of the Site

- Obtain all available information about the movement.
- Study stereo aerial photographs to determine the boundaries of the movement as it may be greater in extent than is obvious.
- Look for similar features on the other side of the ridge and lineations across more distant slopes that may indicate a fault or crush zone. (In crush zones the underlying material has been weakened by fault action and movements, usually earthflows, are deeper and harder to control. Groundwater flow patterns may be less predictable in crush zones and fault lines and in places natural gas may be a complication.)
- Search for any available geological information to determine lithology. The dip slope and strike of strata will be significant in the case of slides.
- How deep is the movement? The height of the head scarp will give some indication but in the case of earthflows may exaggerate the actual depth.
- Locate possible infiltration zones and ponding areas which are often produced by previous movement in the upper parts of slumps and earthflows. (While ponds raise water levels in the surrounding area and presumably increase pore water pressure beneath them, it is dry hollows with substantial catchments that have the greatest impact on infiltration. Particularly in deep slumps in argillite, these hollows concentrate run-off and allowed it to flow to deeper layers through highly permeable crushed rock. As the failure plane is almost certainly of lower permeability, high pore water pressures can develop there, triggering movement.)
- Examine the toe of the slope to see if lack of support there has triggered the movement.
- How active is the movement? (Slumps and slides that have reached equilibrium may be reasonably stable, as long as their toe is not eroded or cut away. Many older farm buildings in hill country are located on the back slope of old slumps, as they may have been the only level sites not prone to flooding. Earthflows cannot be regarded as having long term stability unless extensive stabilisation work has been carried out, and they are almost certainly not suitable as building sites).

10.1.4 Surface drainage

The rotational movement of **slump** failures can form a back-tilted zone, below the head scarp, which has poor surface drainage. Tension cracks and general deformation in this zone allows infiltrating water ready access to the shear plane.

The careful use of surface ditches can drain these ponding areas and reduce infiltration. The number, location and depth of the ditches must be determined for individual sites.

Attention to detail is important to prevent scour erosion of the ditches, particularly in unconsolidated sediments where a surface cover of grass may be more important than a tree cover which leaves the soil bare. Drop structures and flumes may be needed to safely remove drainage water from the movement (those structures have been covered elsewhere in this handbook). As an alternative, willow poles can be planted where scour is a problem. The ditches need to be maintained and cleared of debris regularly to remain operative.

Where dry hollows with significant catchment areas exist, run-off should be diverted from them and out-fall ditches provided. Smoothing of the land surface to shed water may reduce infiltration and a covering of low permeability material such as a clay soil could be considered.

Any part of the slump that has very broken ground may require re-contouring by bulldozer to encourage run-off. Tension cracks should be filled and compacted to reduce infiltration.

Graded banks, very similar in appearance to small farm tracks, may be used to carry run-off to stable land off the slump or slide. The grade must be appropriate for the soil type to prevent scour, but steeper grades remove water more effectively.

Slides and earthflows may have different land forms in the upper regions but the same principles regarding infiltration reduction apply. The shear plane may be more directly exposed in the upper slide and the prevention of infiltration here is critical.

10.1.5 Sub-surface Drainage

This is intended to remove water that has already infiltrated into the soil and so also the lower pore water pressure at the shear plane. Two variations have been commonly used on farm scale works, particularly in the Gisborne-East Coast Region:

Horizontal boring, in which holes of about 50 to 60mm diameter are drilled on a gradient up into the unstable mass for distances up to 30 metres. The holes are lined with perforated pvc pipe of about 25 to 35mm diameter and the water removed is collected by a system of surface polythene pipes and disposed of safely off the movement.

Portable drilling equipment based on a small motorised posthole borer has been constructed for this purpose (see photograph) and the unit can be carried by two people in rough terrain.

The location of the boreholes is more of an art than a science as the method relies on intercepting naturally occurring high permeability zones in the movement. Experience in interpreting landscapes for signs of old tension cracks, shear planes and even fault lines is important. Some remarkably high flows have been recorded such as at a site near Tokomaru Bay in which 3000 litres per hour was removed after heavy rain by 18 bores (Hall J. 1973). Volumes of this size could not be expected from material of uniformly low permeability.

Portable drilling equipment used to install horizontal bore draws. This bore produced an initial flow of approx 3,500L per hour and prevented further slope instabilily. **Photo: Don Miller.**

Spring taps, in which obvious springs on unstable hill slopes are excavated down to parent material to locate the source of the flow. An envelope of polythene, filter fabric and a permeable material (gravel or polystyrene beads) surrounding a length of perforated pipe is buried so as to intercept the flow and take it to the surface. From here it can be piped to safe disposal or used as a stock water supply. It is critical that the open end of the perforated pipe extends to the ground surface on the up-slope side to allow the entry of air to the system. If not, air locks will stop all flow in the pipe.

As the lower sections of an earthflow are generally too mobile for horizontal bores the most effective subsurface drainage will be in the upper sections where groundwater from higher land may be entering the earthflow. Monitoring of a number of earthflows revealed evidence of this.

The best location for spring tap and horizontal bore installations tends to be where ground water seeps to the surface in the upper section of flow. The location of these areas is best done in late summer when vegetation changes are most distinct.

Cut-off drains can be used to intercept groundwater flow, usually, but not necessarily above the unstable part of the slope. They can be particularly useful above unstable sections of farm track as they will improve both slope stability and surface wetness. Location will determined by observation of wet zones on the slope and by the presence of infiltration areas further up-slope.

Surface recontouring of an earthflow north of Gisborne. The areas of recontoured, smoothed and cultivated land contrasts strongly with the areas of untreated earthflow. A graded bank removes water from an area of springs to stable land for safe disposal. Photo: Don Miller.

In general the greater the depth, the greater the possibility of intercepting groundwater, although depth is limited by available equipment and by safety considerations. The most effective cut-off drains will penetrate to a zone of permeable material or fractured rock in which water is flowing. Unstable slopes below ash covered ridge tops could also benefit from extensive use of cut-off drains although expensive. Although untested, this could also reduce the incidence of shallow slips which are common in those circumstances.

Drains consist of a perforated pipe surrounded by a bed of permeable filter material in a back-filled trench. The choice of permeable filter surrounding the pipe is important. While not a problem in fractured hard rock, the force of the escaping water can erode the surface of softer materials and lead to clogged, ineffective drains. For example the use of gravel as a filter in non-cohesive sediments may allow migration of silts to occur. Filter fabrics should be used around the outside of the filter material, rather than directly around the pipe, to maximise flow. Details of filter design can be found in engineering text books (e.g. Cedergren, 1968).

Commercially available filter materials include vertical multi-layered composite filters to collect and remove seepage over several metres depth.

Back-fill material should prevent inflow of surface water unless a second shallow perforated pipe is included for this purpose, as the flow of surface water down to greater depths could cause instability if pipe outlets become blocked. Tree roots, particularly of willows, could block drains if trees are planted too close. If cut-off drains are reasonably straight it may be possible to clear and flush them using a high pressure pump and suitable nozzle, as with horizontal bores.

10.1.6 Surface Re-contouring

Smoothing and re-contouring is aimed at speeding run-off and eliminating hollows that could pond. Earthflows have a more disrupted and broken ground surface than slumps or slides and are generally more shallow. They move in a more regular pattern, being triggered by consistent heavy rain over a number of days (In a Gisborne trial an average of 40mm rain per day over 10 days created approximately as much earthflow activity as Cyclone Bola did with about 400mm in 3 days).

Smoothing and re-contouring the broken surface to reduce infiltration is the most important component of de-watering an earthflow, but this may not be possible until initial surface drainage work has lowered the ground water table. In extreme cases it may take one or two years after ponding areas have been drained by ditches before the land is dry enough for bulldozer operation.

Re-contouring involves the movement of hillock material to larger hollows in the ground surface by bulldozer, to facilitate run-off of surface water. Care should be taken to preserve topsoil for later respreading. Smoothing removes the "hump and hollow" surface characteristics of earthflows. Cultivating and re-sowing pasture follows, and trials have shown that substantial growth of summer feed results, due in part to the shallow ground water table.

10.1.7 Graded Banks

Graded banks, serve to shorten the lengths of run-off overland flow paths, so decreasing the possibility of infiltration. They can also remove the outflow from spring taps and horizontal bores, where these have been installed. These banks are the most vulnerable part of the drained earthflow as water is concentrated there.

Successful earthflow de-watering projects have generally had substantial plantings of willows or poplars along the lower sides of the graded banks where their roots strengthen the soil and compensate for the increased pore water pressure under those banks. A major failure during Cyclone Bola occurred in a dewatered earthflow where the willows were less than 2 years old.

Gully Structures & Pole Planting

Chapter

Chpater - 11

Gully Control Structures

11.1 FLUMES & CHUTES

11.1.1 Description/Purpose

Flumes and chutes are structures that convey stormwater over a gully head or scarp slope to discharge at a point sufficiently downstream so that the plunge pool does not migrate upstream to undermine the gully head. In many situations, flumes are used as a short to medium term measure to control active headward erosion, until associated protection planting has stabilised the gully head.

11.1.2 Application

The design of the flumes and chutes is critical to ensure that they are able to convey peak flows. Flumes were normally designed to convey (at least) a 10 year storm event where the storm duration is equivalent to the time of concentration. Many of the runoff control systems described below are covered in detail by Eyles (1993), using design practices developed by Mr Ian Cairns and other soil conservators in the Central Volcanic Plateau Region. When siting the runoff control structures, alternative flow paths were often incorporated into the design so that when design flows were exceeded, the stormwater flows discharged to alternative outlet points where damage would be minimised as far as practicable. Most failures relating to flumes were due to incorrect sizing for the design storm, or failure at the inlet of the flume by undermining or scouring around sides of the wingwalls.

The design of flumes and chutes has changed over the years. While all flumes and chutes in the past were designed and constructed for specific sites, different products to convey small to medium flows are now available commercially, and can be bought 'off the shelf'. As these products have become available, designs have adapted to incorporate some detention of stormwater above the gully heads with the discharge into the newer types of flumes, chutes or pipes.

When constructing any gully control works where storm water is going to be directed to a pick up point and then safely conveyed past the critical gully head, water should not be directed into the flume or chute until downstream works have been completed. Consider the consequences of a storm half way through the job. It is often a good idea to divert stormwater away from the works site until the project has been completed. Alternatively, if the job is going to take two or three days, wait for a spell of good weather, and then try and complete the downstream part of the job before building the bunds and stopbanks to divert water into the runoff control structure.

11.1.3 Types of Flumes and Chutes

Lip Flumes

These are wooden structures that concentrate the surface flow and discharge it safely over a gully head or scarp slope.

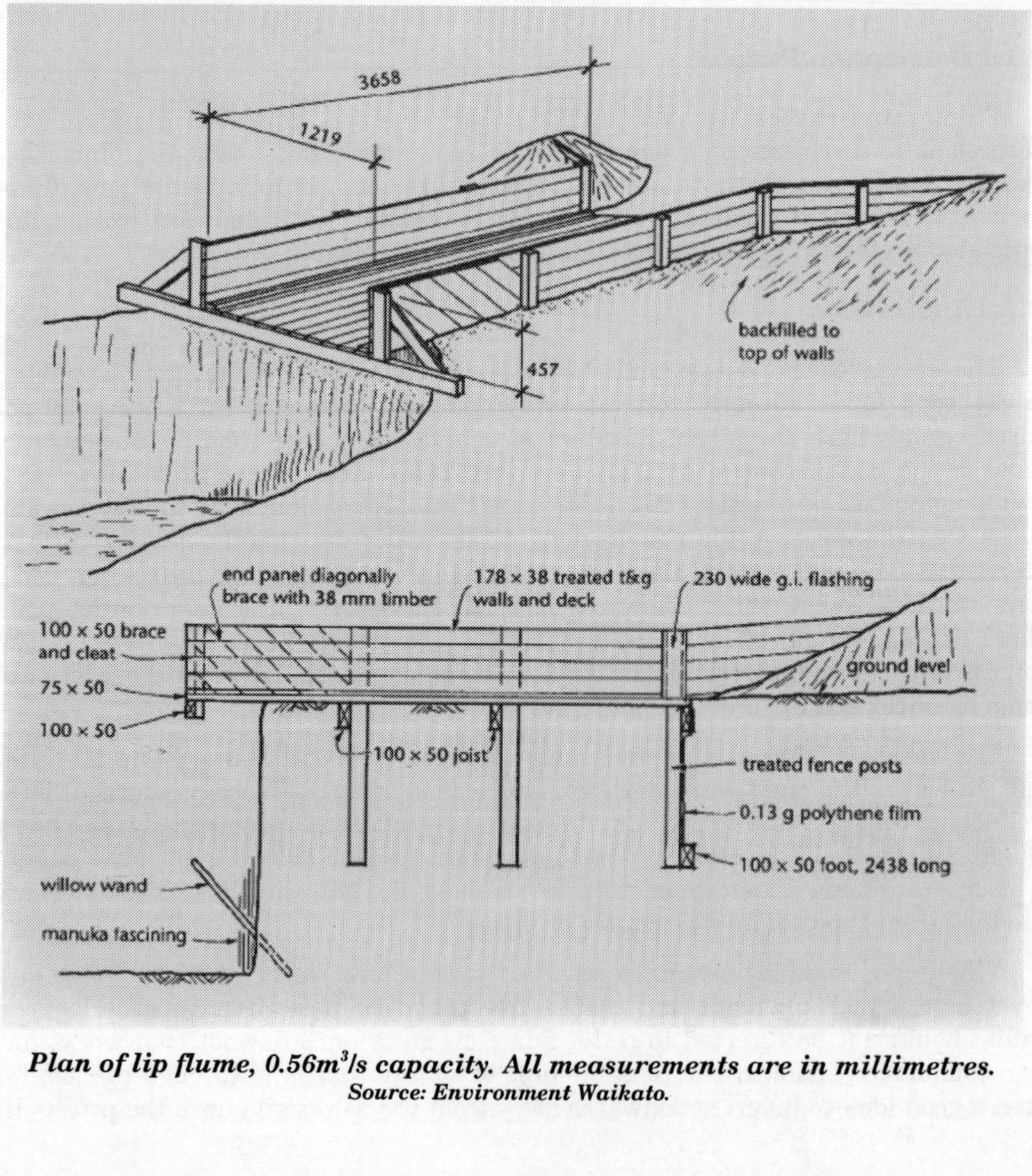

Plan of lip flume, 0.56m³/s capacity. All measurements are in millimetres.
Source: *Environment Waikato.*

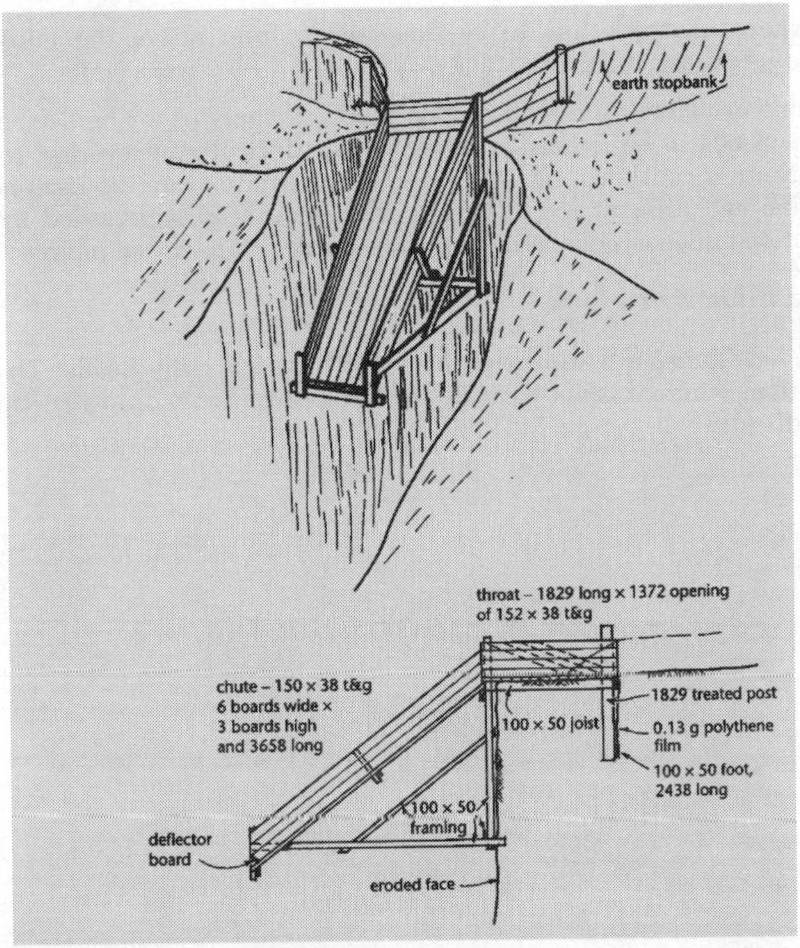

Plan of a typical cantilever flume. All measurements are in millimetres. Source: Environment Waikato.

They need to be constructed using from either ground treated tongue and groove timber or H4 marine plywood. It should be dry, tightly jointed and double nailed with galvanised treaded flathead nails at all bearers and joints. The structure needs to be well footed, and sealed carefully at the inlet using .005mm polythene or equivalent. Wing walls at the inlet need to be well founded and sealed where they join with the sidewalls of the flume. Bunds (stopbanks) controlling the flow of water into the inlet also need to be well founded and compacted.

Cantilever & Escarpment Flumes

Cantilever and escarpment chutes are wooden flume structures designed to be built upslope of a gully head and cantilevered into place.

These structures have the same function as the lip flumes. They are used only in situations where the gully heads are more than 3 metres deep, and often 10 metres or more, or where all work has to be undertaken from above the gully head and cantilevered into place.

The same principles and construction methods apply as for lip flumes. However, construction needs to be lighter to allow for the difficulty of levering the structure into place. Neither cantilever nor escarpment flumes are currently used because they are expensive and difficult to construct. They have been superseded by the use of small detention bunds or ring bank discharging into drainage coil piping.

Suspended Flumes or pipes

Suspended flumes are supported by wires out over gully heads. They are used only in locations where it is not possible to build conventional drop structures because of access, difficulty.

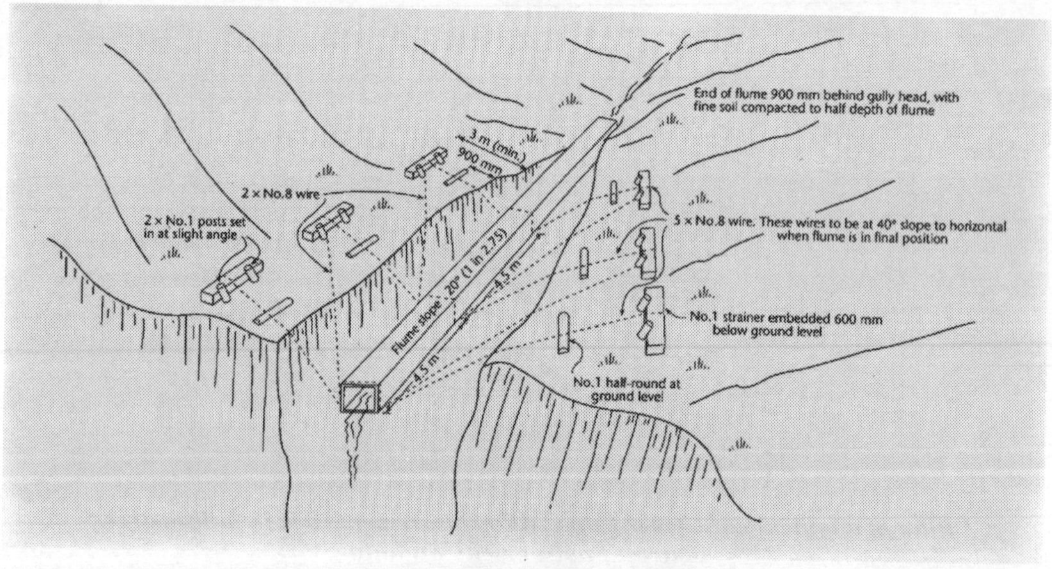

Plan of a suspended flume designed to drop surface flows safely over a 30m head in a difficult hill environment. This was one of a series of measures that included diversion banking to concentrate flows into detention bunds, retirement and planting of the gully head and outflow areas. This flume, constructed in the early 1960s, has disappeared but the gully system remains stable.

The suspended flumes are used to convey the stormwater flows far enough away from the gully head so that the discharge drops and forms a plunge pool that dissipates the energy without migrating back up slope to undermine the structure.

The design of suspended flumes varies depending on site conditions. The wooden constructions are expensive to build and difficult to maintain. The availability of plastic pipes in a range of lengths and diameter has meant that the principle of suspended flumes can be used with pipes as an alternative. However, their use is restricted to a few sites only, and often in conjunction with other practices such as detention bunds.

Wooden Chutes

Wooden chutes are used to convey stormwater flows down steep slopes, or escarpments. They sit on the ground surface and need to be anchored at regular intervals. They discharge at the toe of the slope or escarpment into a dissipating chamber.

Wooden chutes are constructed out of either ground treated timber or H4 marine plywood. The actual designs vary depending on how critical the site may be. They are often used to take the discharge from a culvert pipe safely down a steep slope. While Armco fluming, plastic piping or other products have largely superseded them, they are still used in some locations as they can be constructed on site, and can be sized to accommodate the design flow.

Butyl Rubber Flumes

Butyl rubber flumes are used to convey stormwater down slopes or escarpments, in locations where the slopes are not too steep (less than 20°) and the flume is dug partially into the ground for support. The discharge point needs to be protected to avoid scour problems.

Butyl rubber flumes were used in situations where water control was required for the medium term, and the volumes of water or length of discharge was too great for wooden chutes. They were often manufactured to suit specific jobs and were considered to be a costly option to employ. Butyl rubber fluming is not used often now as there are a wide range of alternative cheaper products that can be used with equal success.

Butyl rubber flume. **Photo: Gisborne DC.**

Corrugated Armco or Steel Flumes

These flumes act as chutes to convey flows safely down steep slopes or escarpments. They may be partially dug into the slope so that they are supported, or can sit on the ground surface and be anchored at regular intervals. The discharge point at the toe of the slope needs to be adequately protected to avoid scouring problems.

These flumes are used very widely, particularly in association with conveyance of runoff from roads and tracks taking discharge from culvert pipes safely down steep slopes below the road.

Alternatively, they can be used in a similar fashion to lip flumes and extend over the edge of a gully head.

11.2 PIPE DROP STRUCTURES

11.2.1 Description/Purpose

Pipe drop structures transport surface stormwater from above a gully head to the gully floor, bypassing the gully head.

Application

Pipe drop structures have largely replaced wooden flumes and chutes, especially where design flows are relatively low. They have a number of advantages over wooden structures:

- They have a lower cost compared to wooden structures;
- They are more versatile and have lower maintenance requirements;
- Inlet systems for pipes are often prefabricated, which reduces the risk of failure;
- More pipes can be added if required;
- They have a degree of flexibility and can move if the gully changes shape.

Pipe drop structures are often used in conjunction with bunds or diversion banks, as well as detention banks or dams. They vary markedly in design and construction, but essentially there are two main types:

- Rigid pipes carrying surface stormwater over or away from the gully head to a safe disposal point;
- Flexible pipes following the gully head or side wall contours down to a safe disposal point. The flexible pipes can be of collapsible form, which expand to take design flows.

All pipe drop structures need to be firmly anchored throughout their length and at the discharge point. When they are running full, they have to cope with the weight of water as well as the energy of water. Pipe structures that drop water to considerable depths will need to cope with high water velocities, and need to be sufficiently robust to withstand the pressure changes. Specialised engineering design may be required to ensure that the structure is sound and well founded.

11.2.2 Types of Pipe Drop Structures

Drop Man Hole Inlet

The drop man hole inlet structure has an inlet into a man hole that allows the surface stormwater to descend safely to the required depth, before discharging out of an outlet pipe to a safe disposal point. The drop man hole inlet requires careful design to ensure that the structure is able to convey the design flow. The design also needs to ensure that the structure is able to cope with the forces exerted by the energy of the water dropping down the man hole, and the high pressure from the head of water.

The structure is commonly used where the gully head is not excessive (less than 3 metres depth) as a substantial amount of earthwork may be required for installation. These structures are used to provide long term protection with relatively low maintenance requirements.

Rigid Pipe Drop Structure

These structures involve the use of rigid pipe either discharging out over the gully head into a plunge pool, or conveying surface stormwater around the gully head and down the side of the gully to a safe disposal point. Where the pipes are run down the slope, they need to be well anchored at regular intervals.

In the past, costly steel or concrete pipes were used for protection of valuable assets. More recently, plastic or PVC pipes have become available, and are more cost effective.

Flexible Pipe Drop Structures

Drainage coil is a non-collapsible flexible pipe e.g. Nova-flo. When used with a bund or ring bank around the top of the gully head, it has proven very successful for treating small gullies (less than 5 metres high). The drainage coil is used to convey surface stormwater around the gully head to a safe disposal point on the gully floor.

The drainage coil pipe used should always be the non-perforated pipe that does not have drainage holes in the pipe wall. The pipe should be either anchored or dug in to ensure that it is stable. Inlet systems vary in design, but often incorporate a small drop man hole buried so that the inlet is at ground level. Alternatively, a simple culvert inlet with a small head wall can be used to drain stormwater from diversion bunds or detention banks. For remote sites, sandbags filled with a dry cement/soil mix can be used to construct small inlet headwalls. Sandbags should be filled to no more than one third full. The ends are folded back to seal the sandbag, and the headwall is constructed using the sandbags in a manner similar to laying bricks, but allowing for a slight batter slope on the headwall. Following rain, the headwall hardens and becomes a permanent structure.

Collapsible tubing such as Pyvac, Structure-flex, or UV stable lay-flat piping can be laid to remove surface stormwater from gully heads. The collapsible tubing can be factory prepared to cope with specified volume of runoff, and is sufficiently flexible to follow the contour of the scarp face. Operates in a similar fashion to drainage coil pipe, in that it is often used in conjunction with a ring bank or diversion bund that protects the top of the gully head, and directs water into the pipe inlet.

The collapsible tubing needs to be well anchored at regular intervals to prevent the pipe from twisting and blocking. Also, the collapsible pipe needs to be firmly attached to the inlet pipe as the weight of water over the gully drop off can exert enough force to split or break the material. Pyvac tubing is produced in rolls 30 metres by 1500 millimetres with a full width fitting over a 457 mm diameter pipe. Sheets can be welded to fit larger pipes.

Collapsible tubing has been largely superseded by drainage coil pipe because drainage coil is cheaper, more versatile and requires less maintenance.

11.3 SINK HOLES

11.3.1 Description/Purpose

A sink hole is a drill hole usually with a diameter of 1-1.2 m, and a depth in excess of 6 m, filled with rock rubble. It is located immediately upslope from a bund on low angle terraces in an ephemeral drainage line. The practice is similar to normal soak holes used for the discharge of stormwater from houses, except the soak hole is extremely deep.

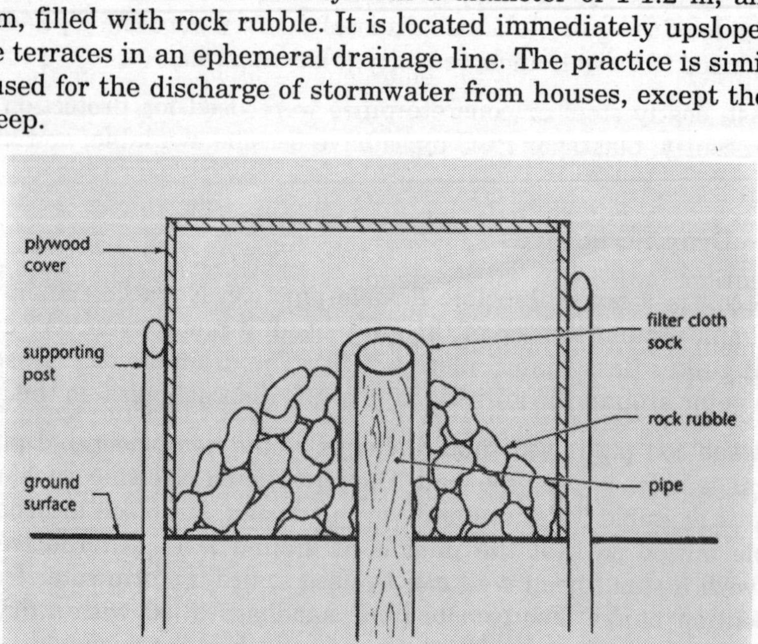

Sink hole cover for heavily stocked locations. ***Source: Environment Waikato.***

Application

The downstream sink hole and bund should be located no closer than 150 metres from the gully head or terrace edge to ensure that groundwater recharge via the sink hole does not weaken the scarp face and cause collapse. The holes are normally used in series to provide for continuous drainage directly into the ground. Their use has been confined to deep gravel/sands river terraces in the Waikato region. The inlet system should be fenced to prevent stock and machinery from going too close to the structure. Careful attention should be given to ensuring an efficient filter system is put in place to prevent sealing and failure of the system. The practice is described in detail by Eyles (1993).

11.4 DETENTION BUNDS/DAMS

11.4.1 Description/Purpose

Detention bunds are up to 1 m high with a 150mm or similar-sized discharge pipe through the base. They are usually used in series along a valley floor or upslope from a gully head and are often used when limited storage is required.

Detention dams are larger structures up to 2.5 m to 3m high and provide a greater storage volume than detention bunds. They are often located close enough to the gully head so that the drainage coil discharge pipe is able to convey water over the gully head to a safe disposal point on the gully floor. Sometimes two or three detention dams may be constructed in series down an ephemeral valley, with the downstream structure conveying stormwater over the scarp face/gully head.

There is provision made for an emergency spillway around the side of the structure (at 2 m height), with the overflow directed to a safe alternative outlet as far as practicable.

Application

Both detention bunds and dams detain stormwater so that flood peaks are attenuated. They are normally used in conjunction with other works involving fencing, retirement from grazing and planting of the gully head and associated steep escarpment area. The detention structures are normally designed to provide for a minimum of 100 cubic metres of storage for each hectare of contributing catchment. The method is used primarily for small catchments of up to 40 hectares only.

Care needs to be taken with both the detention bunds and dams, so that discharge from the outlet pipe does not cause scour.

The detention bunds and dams can only be used if the site is appropriate, and they provide protection for small to medium storms (5 to 10 year return period) if used in conjunction with other works as described above. The practices are described in detail by Eyles (1993) and Environment BOP Fact sheet 10/98.

11.5 DIVERSION BANKS = DIVERSION BUNDS ON EARTHWORKS

Description/Purpose

Diversion banks are earth banks that guide surface flows across a slope to a safe or controlled outlet. They are used either to protect a gully head, by diverting surface stormwater flows to a safe outlet, or to collect surface stormwater and move it to a particular spot for safe disposal or detention.

Application

Diversion banks are often used in conjunction with a controlled outlet pipe or flume system. They should be designed for the 1 in 20 year storm event and constructed on a grade that does not exceed the scour velocity of the particular soils on site. This may be as low as 1.25%, so a survey grade line will be required. Diversion banks need to be well compacted and grassed as soon as possible following completion. If the batters are not too steep, the diversion bank will blend in with a paddock and become part of the landscape.

This practice is described in more detail by Eyles (1993) and is similar to the diversion channels/bunds described in the chapter on earthworks.

11.6 GRADED WATERWAYS=GRASSED WATERWAYS IN PASTURE & CROPLAND

11.6.1 Description/Purpose

A graded waterway is a mechanically constructed grassed waterway that intercepts and transports surface runoff to a safe outlet. The function of a graded waterway is to regain the natural drainage grade in a gullying drainage system, enabling flows to be managed without scour or further gully formation.

Application

As with diversion banks, it is essential that the waterway grade keeps runoff to less than the scour velocity. All graded waterways should be well compacted, and grassed as soon as possible after construction. Alternatively, natural stable ground can be used.

Maximum erosion proof grades of waterways depends on factors such as the design flow, the shape of the waterway, frequency of peak flow etc. Instead of slope angles, maximum flow velocities are generally recommended beyond which erosion of channel will start to occur. For grassed channels, a conservative estimate of this velocity is about 2m/s (ref: Hydraulic Design manual, Humes Industries Ltd). This value will vary depending on the types and state of the vegetation, e.g. type of vegetation, height, density, stiffness etc.

The following table indicates acceptable velocities of flow within various grass lined channels on a range of slopes.

Table 11.1 Permissible Velocities for Grass Lined Channels

Channel slope	Lining	Velocity* (m/s)
0-5%	Grass	1.25
	Kikuyu	1.85
5-10%	Grass	0.81
	Kikuyu	1.5
>10%	Grass	n/a
	Kikuyu	1.25

*For highly erodible soils decrease permissible velocities by 25%
(Modified from Soil and Water Conservation Engineering; Schwab et al).

Simple engineering formulas can be used to work out the design flow and channel size and shape. Channels that will have velocities greater than about 2 m/s may need to be reinforced or armoured (using fabric, rock etc).

The graded waterways can be installed in situations where the eroding gully head is not excessively high. The site needs to be excavated to provide a firm foundation prior to filling to correct grade. The fill material should be formed in layers and firmly compacted as filling proceeds. Different geotextiles are now available with which to reinforce and grass the graded waterways. Reinforced waterways can withstand significantly higher velocities i.e. can be installed at steeper grades without erosion occurring.

Graded waterways are suitable for grassed spillways at detention dam sites. Care should be taken when grazing stock on grassed waterways, as any exposure of the ground surface can initiate further erosion.

11.7 DROP STRUCTURES

11.7.1 Description/Purpose

A range of drop structures are used to convey surface water safely over a gully head. Most of the structures are capable of controlling a gully head drop of up to 3 metres in height. However, many are only used on short gully head drops of 1 to 1.5 metres.

Application

While many of the drop structures are expensive to construct, they are also relatively permanent. Generally there is less reliance on planting for long term stabilisation. Some of the drop structures described below can also be used in permanent streams.

All drop structures should be carefully designed to ensure that they are able to convey the design flow. Provision needs to be made for storm flows in excess of the design storm, when the structure is overtopped. Drop structures are designed to be site specific. Sometimes, the ideal structure may be a combination of two or more of the generic structures outlined below.

Drop structures generally have two components:

- The 'drop' section where the water is conveyed from one level down to a lower level; and
- The 'energy dissipator' which may be a plunge pool or a flexible structure such as a tyre mattress.

The wide range of construction materials and geotextiles now available means that many different types and combinations of drop structure can be built.

11.7.2 Types of Drop Structure

Concrete Drop Structures

These are constructed entirely of concrete, concentrate the surface stormwater at the top of the gully head, and discharge the water over a constructed chute to dissipate safely at the toe.

Concrete drop structure – spillway over dam. *Photo: horizons.mw.*

The inlet needs to carefully designed and constructed so that the structure is not outflanked. These types of drop structure are expensive, but will last for decades. They have been used to protect valuable assets and are often used as spillways for dams. The structure requires careful engineering design to ensure that the design storm can be conveyed, and that the energy is dissipated at the toe of the structure.

Geotextile/Interlocking Concrete Block Structures:

These structures are similar in principle to the concrete drop structures except they are built using geotextile cloth with interlocking concrete blocks laid on the surface of the cloth to hold it in place.

Geotextile interlocking concrete blocks. **Photo: Wellington RC.**

Sheet piling drop structures. **Photo: Gisborne DC.**

The structure is designed to be site specific and is built completely on site.

This type of structure is expensive, but effective and permanent. The site needs to be carefully prepared with experienced staff carrying out the construction.

Rock Rip-rap Drop Structures

These drop structures use rock rip-rap to convey surface stormwater from the top of a gully head safely down to the bottom of the drop. The median diameter size of the rock rip-rap should be calculated to ensure that it is able to lock together and withstand the forces generated by the stormwater.

These structures have proven to be successful, as the rip-rap provides a structure which is flexible, and can adjust to any changes in the shape of the gully head over time. The inlet to the drop structure is often formed from gabion mattress or concrete to ensure it is not outflanked. The rip-rap may need to be topped up after a few years, as the structure settles. Geotextile cloth should always be laid underneath the rip-rap. During construction, the channel slope is shaped first. Geotextile cloth is then laid down, the inlet constructed, and rock rip-rap placed. Depending on the site, the inlet may be constructed last.

Rock gabion drop structure. Photo: Wellington RC.

Sheet Piling Drop Structures

Sheet piling is sometimes used to control small gully heads (up to 1.5 metres in height) particularly in difficult situations such as permanently flowing streams. Sheet piles are lengths of channelled steel that are driven in as piles but interlock so that the completed structure forms a continuous wall. The sheet piling is driven (or vibrated) into position with heavy machinery and provides a permanent gradient control.

The sheet piling needs to be long enough so that the plunge pool formed at the toe of the drop structure does not undermine the sheet pile walls. This method is an option that can be used for river training works.

Rock Gabion Drop Structures

Rock gabion drop structures are normally used to control small gully heads up to 1 metre in height. They can be used in both ephemeral gullies and permanently flowing watercourses. The structure comprises a rock mattress (rock inside a wire mesh) that forms the downstream energy dissipator, with a rock basket on the upstream end. The stormwater flows over the rock basket to land on the rock mattress. The rock mattress needs to be long enough, that any plunge pool forming at its toe does not undermine the structure. Engineering design will be necessary to ensure that the structure works.

These structures will require machine work to construct. They should be carefully designed so that the size and weight of rock used in the structure is able to stay in place. The mattress is flexible and will drop if a plunge pool starts to form at the downstream end. Also, it is important that the rock basket and sides of the rock mattress are carried up the sides of the channel and fitted into the natural ground contours in a hydraulically smooth fashion. Geotextile cloth should be laid underneath the structure to ensure that fine material is not scoured out. When these structures are used in rocky streams, tyres are sometimes wired onto the top of the mattress to protect the wire mesh from being broken by the moving rock bedload during major floods.

Normally these structures have a life of 15 to 25 years, and are supported by planting.

Boulder/Tyre Drop Structures

These structures are similar in principle to the rock gabion/mattress drop structures except they are constructed out of boulders and tyres, with the tyres wired together and laid over the downstream part of the structure. They are normally used in permanently flowing streams where a substantial structure is required to control channel gullying.

These structures have been used in the Wairarapa by the Wellington Regional Council and require experienced people to ensure they are designed and constructed properly.

Geotextile Reinforced Grass Drop Structures:

These structures are used in ephemeral gullies to safely convey the surface stormwater over a gully head or steep slope. They comprise a shaped grass chute that has been reinforced by geotextile so that the roots of the grass are bound together and are able to resist normal scour velocities during flood flows. The site is shaped and the geotextile is laid down. Topsoil is then spread over the geotextile and grassed. The grass should be firmly established before the drop structure is used. Some form of energy dissipater such as rock or gabion basket is required at the downstream end of the structure.

These structures have only become possible with the advent of a range of geotextiles in recent years. The appropriate geotextiles should be chosen following careful design to ensure that the velocities will not scour out the structure.

11.8 DEBRIS DAMS

11.8.1 Description/Purpose

Debris dams are structural controls that are built in the floors of eroding gullies. Their purpose is to stabilise the gully floor so that trees can be established to stabilise the gully sides. While trees are the main long term tool for V-shaped gully control, they can be difficult to establish if water channels continually undermine the toes of the hill slopes.

Application

Debris dams are largely confined to the control of V shaped gullies. They are normally built in series over time, with the base of the upstream debris dam level with the top of the debris dam below. However, locating a suitable site to commence debris dam construction is an important part of ensuring their success. It is important that the site is able to give sufficient support to the sides of the dam.

Debris dams can be used for a number of purposes (Gair 1973):

- To provide a stable area in an eroding gully, thus facilitating the establishment of trees or other vegetative control measures;
- To effect grade control, by eliminating bed level fluctuations;
- To reduce channel slope angle;
- To raise bed height, thereby supporting the base of gully walls;
- To increase bed width, and with it reduce water velocity;
- To centralise water flow in the channel;
- To trap and hold debris. This not only gives associated tree planting a better growing medium, but also helps reduce deposition downstream.

The height of debris dams (where the water flows over the centre of the structure) should be approximately 600mm on completion of construction. Following the formation of a plunge pool, this height may increase to 750800mm. It is important that debris dams are not built higher than this, as they will become prone to failure. Over the years, many different types of debris dam design have been built. A common factor in all types of construction is the need for the structure to be stable when the downstream scour hole has formed.

Timber debris dams. ***Photo: Gisborne DC.***

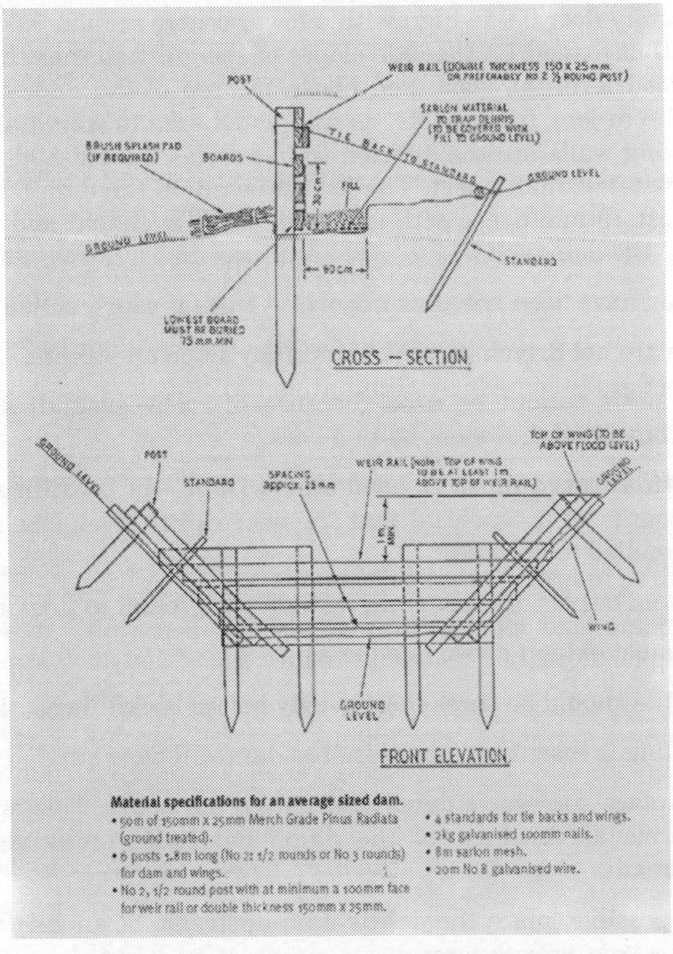

Timber dam construction guidelines.

In the Gisborne – East Coast region of the North Island, early dam construction made use of manuka brush, which was often in plentiful supply. The brush was used for fascines, acting as an energy dissipator below the debris dams. The most successful design was the Timber Debris Dam. The Timber Debris Dam was based on an earlier Pole and Netting Dam designed by J Gair, but used ground treated timber as suggested by H Pearce. OM Borlase of the Poverty Bay Catchment Board, and WR Howie, Water and Soil Division, Ministry of Works undertook early design work on debris dams. During the 1960s and 1970s, thousands of debris dams were constructed, particularly in the Gisborne / East Coast area. However, very few have been built since loss of Government subsidies for soil conservation work, in 1988.

11.8.2 Types of Debris Dams

Timber Debris Dam

The timber debris dam is built out of ground treated timber and comprises a dam wall across the gully floor 0.9 m high with a lower centre section 0.6 m high. The dam has wing walls that extend up the side slopes of the gully so that the top of the wing wall is above flood level (at least 1 m above the top of the dam face). The dam is anchored with 1.8 m long fence posts, as well astiebacks to warratah standards. The dam wall and wing walls are constructed from ground treated 150mm x 25mm rails spaced approximately 25mm apart. The lowest board on the dam face is dug into the gully floor at least 75 mm deep, with geotextile cloth anchored and laid part way up the inside face of the dam wall.

The following have been noted as important tips for construction:

- If posts are not driven in, make sure they slope up-stream.
- Where posts cannot be used (because of rocky ground) standards can be used. They must, however, be tied back.
- The bottom board must be level otherwise it will be difficult to line up the other boards. It is essential that the weir rail is level, or else the water will be channelled to one side.
- All bottom boards must be hard against the ground to prevent undercutting.
- Wings must extend above flood level.
- Geotextile should be spread out evenly before backfilling.
- Backfilling is essential; otherwise the dam will blow out.
- Do not place successive dams closer than 3m apart. This is because a scour pool forms at the base of the dam, and could undermine another dam downstream.
- Where possible, place the debris dam upstream of an existing strong point, such as a tree, or rock face.

Pole and Netting Debris Dams

The pole and netting debris dam was designed in the 1960s, by trying different designs and modifying them over time. The pole and netting dam is similar to the timber debris dam except it is constructed using netting, warratah standards, polythene and manuka poles. The manuka poles were used for the rails on the wing walls and weir rail. If manuka was not available, polar or willow poles were used instead.

The pole and netting dam was largely superseded by the timber debris dam in the 1970s.

Brush Debris Dams

Brush debris dams were comprised of manuka brush laid 0.5-0.6m thick across the gully floor after rough shaping of the channel. The brush is laid so that the ends overlap, in alternate layers with butts upstream, then butts downstream. Two rows of standards are then driven through the brush in a line across the channel. No.8 wire is then threaded through the top of the warratahs and tightened to pull the brush down firmly. The weir is formed from the brush material with two wings up the sides of the gully. At least 12 poplar or willow poles were planted on the edge of the structure in a pair pole planting pattern.

By the 1970s brush dams were seldom used. However, they did prove useful where there was some degradation in the gully floor, but the sides of the gully were still stable. The brush acted in a similar manner to mulch and helped prevent the mudstone gully floor from drying out and flaking. The dams were cheap, trapped moving bedload (rubble) efficiently, and the manuka brush lasted well. However, the dams did not cope with any erosion of the valley sides, and were not used if brush was not readily available to the site.

'Bag' Debris Dams

The 'bag' debris dam was developed to help collect debris as it moved down the gully, and the collected debris was used as an energy dissipater. The dam was comprised of angle iron either bolted/wired together or driven in a line across the gully floor. Wire netting mesh was then attached to the structure and folded back so that it formed a bag downstream of the angle iron that collected debris. As the bag filled, it tended to settle and act as a mattress downstream of the supporting structure.

In the Manawatu and Rangitikei-Wanganui areas, this technique was developed and used successfully. In other areas, where there was more risk of lateral movement, they required a lot of maintenance and were not used extensively.

Log and Slab Dams

Some of the first dams used boulders, logs or other debris already trapped in the gully floor. These later evolved into using split logs (slabs) or poles placed vertically across the gully floor or horizontally across the gully with lateral support.

'Spider' Dams

Some dams used pre-fabricated angle irons that were then placed as a debris dam structure across the gully floor. Ranges of different materials were used to form a mattress below the dam.

Various Designs

Various other designs were used throughout New Zealand, incorporating aspects of the designs outlined above. Often, the designs were "one-off". At times, they were constructed off-site, then brought in and installed in the gully.

11.9 SURFACE PROTECTION IN GULLIES

Surface protection is occasionally used for gully control. The appropriate techniques are described in Section 16.2 (Mulches, coatings and tackifiers). This page merely gives a few additional observations specific to their use in gullies.

11.9.1 General

Surface protection in gullies can involve grassing (the most common measure), mulching, fabric and all manner of coating substances such as aggregate, chemicals. Surface protection helps to protect the soil against raindrop, sheet and rill erosion, which can aggravate existing gullies. (Also see Earthworks – stabilisation)

Surface protection will not control erosion from concentrated flows, so appropriate runoff control measures should be applied first. Vegetative surface protection such as grassing or hydroseeding will establish more successfully if there is adequate moisture, suitable soil medium, and fertiliser. Non vegetative surface protection such as mulches, or geotextile fabrics will give instant cover, but may need follow up maintenance or establishment of vegetation in the longer term. Some surface protection measures (such as, geotextile fabrics, aggregate etc) can be applied to the flow paths of surface stormwater runoff.

Types of Surface Protection used in gullies:

- Grassing
- Hydroseeding
- Mulching
- Geotextile Cloth
- Erosion Control Blankets
- Aggregate

11.10 MECHANICAL INFILLING OFGULLIES

Description/Purpose

This involves the contouring of the ground surface with earthmoving machinery to infill small U shaped gullies or tunnel gullies. It is carried out in association with surface stormwater runoff control and follow-up surface revegetation. The gully infilling is normally part of the surface contouring to change drainage patterns, and provides for a more productive land use on eroded ground without any productive potential.

Application

Mechanical infilling of gullies is an expensive option that can be undertaken to address small U shaped gullies, and tunnel gullying problems. Mechanical infilling is usually carried out in conjunction with reconstructing drainage patterns, and establishing stabilising ground vegetation. Where small U shaped gullies occur on short terrace faces, gully infilling, contouring of the face to an easier grade, controlling stormwater runoff to reduce velocities, and follow up topsoil and grassing can control the active gully erosion and substantially reduce the risk of further gullying.

Infilling of severe tunnel gullies was carried out at the Wither Hills, Marlborough with diversion banks installed to direct water back to ridges, and follow-up planting and surface revegetation.

Mechanical infilling of gullies is not carried out to any great degree in New Zealand because of the large costs involved. However, following infilling operations at Wither Hills, studies have shown that although the costs are high, increased returns from production alone following treatment can pay for the work over a period of time. Furthermore, increased production costs do not take into account other benefits such as easier management, more versatile land use etc.

Mechanical infilling of gullies should only be considered on small scale gully erosion problems. The costs for infilling large gullies will generally be prohibitively expensive, and potentially prone to failure.

Chapter - 12

Pole Planting

12.1 DESCRIPTION/PURPOSE

Tree planting is the traditional long-term control measure for gully erosion in many parts of New Zealand. In particular, tree planting is used (in conjunction with other measures) for the control of V shaped gullies. The main purpose of planting is to stabilise the sides of gullies. However, planting also stabilises the gully floor in the long term as tree roots extend beneath the channel. Well planned planting stabilises the soil through the binding of roots, as well as dropping soil pore water pressure by evapotranspiration. Planting of trees in a gully's catchment will also help to modify and attenuate peak runoff for small to medium sized storms.

12.2 APPLICATION

Normally specific planting for gully control is confined to willows and poplar species only. Specialised planting may be used in particular regions e.g. Erythrina in Northland or native planting adjacent to ecologically important areas. Willows and poplars are normally planted as stakes or poles for specific gully control works where there is a potential gully erosion problem. This is carried out by pair planting or staggered planting along either side of the gully floor or eroding channel. Alternatively, poles can be space planted on the sides of gullies to help control associated erosion which can aggravate the gully erosion. Willow poles are also used for control of tunnel gully erosion. Extra long poles (up to 5 m long) may be necessary in some places and these are planted directly in the holes formed by the tunnel gully erosion, or by pair planting if the tunnel has collapsed. Polar poles may also be used, although willow poles perform better on wetter sites. The poles should have protectors fitted. Poplars and willows are ideally suited to gully control as they have a strong fibrous root system that helps to stabilise gully floors., Because they are deciduous, they do not shade out grasses and other ground cover, providing they are not planted too close to each other. Initial pair pole plantings of poplar or willows at 5 or 8 m spacing, has proven to be too close after two or three decades. Ideally pair pole planting should have final spacings of 10 to 15 m between pairs. On very small gullies, staggered planting is preferable to pair pole planting.

If willows or poplars are planted out as stakes, the plantings will need to be protected from stock.

12.3 PRACTICE

This involves the planting of unrooted poplar and willow poles into land which is susceptible to slips. The aim is to have a surface spread of roots through the ground which holds the soil together while also drawing water from the ground. Pole length can range from 1 – 305 metres, depending on whether stock are excluded while they establish.

Poles, like all living plants require careful handling and care must be taken to prevent them drying out prior to planting. Studies show that standing the poles in fresh water for 48 hours immediately after cutting will improve their establishment in the field. Initial root growth draws on moisture reserves in the pole. In some areas it is the practice to expose fresh wood at the base of the pole shortly before planting. This is done by pointing the pole with a sharp cutting tool prior to planting.

12.1.1 Pole dimensions

Pole length and small end diameter (SED) will vary according to site conditions, where they are to be planted, and the class of animal that will be grazing the paddock.

All animals will rub against poles as well as chewing the bark where it is not protected, so a bigger stronger pole, well rammed into the ground, will resist movement better than lighter ones.

Table 12.1. Classes of pole (giving typical dimensions)

Classes of Pole	Length	SED
Cattle	2.5 - 3.5 m	25 mm
Sheep	2.0-2.5 m	25 mm
Retirement pole (stake)	1.0-2.0 m	15 to 20 mm

12.3.2 Planting poles

Poles should be in the ground before shoots start to appear, generally before the end of August, although some species have a later bud break.

There are many ways to plant poles, but the most effective ways are driving or digging. For driving, the pole big end needs to sharpened using an axe or slasher. In some areas poles are slice cut in the nursery which eliminates the need to point the pole on site. If the ground is dry and hard, a pilot hole is made with a crowbar, the sharpened end is placed in the hole and then driven into the ground not less than 60cm, with a pole driver. This is a length of pipe, larger in diameter than the pole, about 800mm long and fitted with two handles. A spike may be fixed to the driver in order to make a pilot hole. If the ground is moist and soft, poles can be driven without the need for a pilot hole.

Widespread gully (above) and spaced (below) planting of poplars to control erosion on mudstone soils, Kanakanaia Valley. *Photo: Don Miller.*

When digging a hole, use a spade to dig a hole 60cm deep. Place the pole in the centre of the hole and ram the earth with a rammer from the bottom to the top, taking care to avoid damaging the bark.

It is necessary for the pole to be firm in the ground while roots are developing. Any movement caused by animals rubbing or by wind will cause the roots to break. If a space develops between the bark and the earth the pole will dry out. In some soils, poles may work loose as the soil dries and shrinks. Re ramming may be necessary and farmers should be made aware of this requirement.

12.3.3 Pole placement and spacing

Poles will struggle to grow in thin or dry soil. So sites where there is moisture, and a sufficient depth of soil to allow the pole to penetrate 60cm are essential. Poles will eventually grow into large trees, so spacing should be carefully considered. Trials have shown that a 10m spacing will develop a full root mat between trees at five to ten years age. Between ten and twenty years age, roots will spread more than five metres from the trunk, so the stand may be thinned by removing every second tree i.e. a final spacing of 20m. However this should not be done on particularly unstable sites, where root density at five to ten metres radius may be insufficient to bind the soil.

Poles need only be planted on the unstable parts of a slope, identifiable by micro-relief which indicates past failure (regrasses slip scars and debris hummocks). So actual planting density in a paddock will vary, and average out at less than the theoretical 100 poles per hectare. Establishing a pole in a harsh environment will result in losses, so closer spacings may be required initially, with thinning carried out later. Alternatively, plant replacement poles in the gaps.

Willow poles form a fibrous root mat and do better in wet conditions, such as along stream beds and gully floors. Poplar poles are better on drier sites, although large leafed varieties should not be planted in windy sites. Poles are planted to prevent erosion, so it is preferable to site them where erosion could occur sometime in the future. Planting them in eroded sites (which is common) is worth doing, as it will stabilise remaining subsoil. However, it is much harder to establish poles once the damage has been done.

12.3.4 Pole protection

All poles that are planted in paddocks where animals will graze must have protectors fitted to them. Ideally cattle should be kept out of the paddocks where the poles have been planted for 18 months to 2 years to allow the poles to establish. Other animals such as goats and deer should also be kept out for the same length of time. In practice this rarely happens, so protectors are fitted instead.

Pole protectors, or plastic sleeves, are commonly used where grazing animals are presents. There are two types, 'Netlon' and 'Dynex' and both are slid over the poles at the time of planting. They are 1.7m long for use with cattle poles and slightly shorter for sheep poles. Netlon sleeves need to be stapled at the bottom to the pole.

If stock can be excluded for two years, 12m stakes may be planted, a substantial cost saving, compared with the larger poles.

12.3.5 Other benefits

Willows and poplars are deciduous, and this is a great benefit to pasture and animal production. Firstly the trees will provide shade for animals from spring to autumn. Then, in winter the trees have shed their leaves, so pasture growth is not suppressed. Even in summer, enough light filters through the trees for pasture to keep growing. Field trials have demonstrated that annual dry matter production is depressed by about 15-25% beneath growing trees (Gilchrist et al 1993, McElwee 1998, Parminter and Dodds pers. Comms.). The canopy shading loss is roughly counterbalanced by pasture growth between the trees, on ground that otherwise would have been lost to erosion (Gilchrist et al 1993, Hicks et al 1995, Hicks pers. Comm.). The leaves are high in nutrient and are excellent food value in times of drought or other feed shortages, dropping about 500kg of dry matter a hectare in autumn.

Poplars are recognised in other countries as having a high value for timber, fibre and match production. In New Zealand this value has not really been appreciated by growers or industry. Unpruned and inaccessible trees are of no value to anyone, including the farmer. Occasional silviculture is recommended to:

- prevent trees from splitting and falling (which can damage fences and block tracks)
- maintain good growth form (which minimises pasture suppression by the canopy)
- ensure a defect-free trunk, if harvest for timber is envisaged.

Specific recommendations for matching poplar and willow cultivar to sites, planting techniques, and silviculture, are given in a series of pamphlets recently published by HBRC and TRC.

12.4 REFERENCES

Hathaway, R. & Van Kraayenoord, C, (eds) 1987. Plant Materials Handbook, Water and Soil Division MWD

Wilkinson, A. G. FRI Bulletin No. 124, 17 The poplars, Forest Research Institute 2000

Hawkes Bay Regional Council, Environment Topics – Planting Poplars and Willows 1995.

Forestry & Shelterbelts

Chapter - 13

Erosion Control Forestry

13.1 EARTHFLOW STABILISATION WITH EROSION CONTROL POLE PLANTING

13.1.1 Shallow Earthflows

The trigger for renewed movement of an established earthflow is an excessive amount of infiltrating water causing swelling of the clay and raising pore water pressure. Control is obtained by either reinforcing the soil with tree roots or by reducing the pore water pressure (either by de-watering or evapotranspiration by trees).

Tree species that will establish from poles are almost without exception deciduous and so soil moisture removal during the winter months will be virtually zero. There will certainly be moisture removed when they are in leaf but the value of the moisture deficit that will be carried into winter has not as yet been adequately established. The stabilising impact of deciduous trees is mainly through the reinforcing effect of their roots and it is necessary that they be planted at close spacings.

The depth of penetration of tree roots will have a significant impact on their effectiveness but may be limited by soil conditions. The lower surface of the shear plane may have fine grained material and water may be perched there, both conditions that will restrict root growth.

Planting patterns: Five approaches have been used successfully:

- Very close planting of Salix matsudana on a regular 3 metre by 3 metre spacing with no surface drainage. With 7 year old trees this successfully held a small, previously active, earthflow near Gisborne during cyclone Bola. The only other earthflow at that site that did not fail was afforested with P radiata.
- A wider spacing (approx. 6 metre by 6 metre) of poplars and willows planted after a long period of de-watering. A very active earthflow affected area of bentonite clay, planted in this fashion survived 600mm of rain in 3 days during Cyclone Bola (photos) with minimal damage, but the works were very expensive to install.
- Willow poles closely planted in rows adjacent to graded drainage banks in association with a comprehensive dewatering program. Poles were also

planted wherever scour could have been a problem in other surface drains. The works, on a number of properties, were spread over several years to allow gradual drying out of the saturated soil. Both slope stability and pasture production were greatly improved.

- Planting of poles in pasture on a wide spacing (greater than 10 metres) with no de-watering. These generally reduce erosion by 50% or more compared with unplanted ground. A key factor may be the absence of exceptionally wet winters in the following 5 to 7 years, which could give the trees a chance to establish, and to reinforce the soil mass. This lower cost method may be most effective where earthflow movement is sporadic.
- The planting of small gullies within the earthflow, may prevent worsening earthflow movement as gully enlargement is a major destabilising factor.

13.1.2 Deep Earthflow

The forces involved in deep seated movements are such that even if the roots of trees could penetrate to the shear plane the relative proportion of shear strength they could provide would be minimal. Stabilisation of deep-seated movements relies on de-watering and the most effective trees to lower the water table will be fast growing evergreen species.

If poles of deciduous species are close planted (within a 10 metre spacing) the general increase in shear strength of the soil mass may have a slight benefit, but a combination of de-watering and afforestation will give the most effective low cost treatment.

Full stabilisation of a deep-seated movement, even if it is possible, will usually involve a full civil engineering investigation and subsequent works.

13.2 EROSION CONTROL FORESTRY FOR GULLY STABILISATION

Large scale planting or afforestation with Pinus radiata, Eucalyptus or other species is often used for catchment planting. However, pines should not be used too close to eroding gully channels as they do not have the secondary fine root system necessary for gully control, and they will shade out the willows and poplars. Normal distances back from the eroding gully is at least the height of a mature pine tree. If production trees such as pines are planted, consideration should be given to proposed harvesting options in the future, and whether planting boundaries are appropriate to allow for sustainable production forestry. Trees are normally planted out as open rooted seedlings, and stock are excluded from the planted area. Seedling trees can be planted out with protectors, but careful grazing management is required, and the cost is higher.

Planting for long term control on U shaped gullies is only carried out if the gully head is not too deep (less than 2 m). Normally, planting is undertaken on U shaped gullies to provide a vegetative regime that is easy to maintain, or to provide an income when trees are harvested. Normally, willows and native species are planted within the purely protection areas, while other species such as pines, or special purpose production species are planted in areas that can be safely harvested without compromising the gully control works.

Most U shaped gullies are fenced and permanently retired from grazing, because the gullies are a hazard for stock. It is advisable to provide a gate or rails (preferably wired shut) in a retirement fence to make it easy to get stock out of the retirement area if they do manage to find their way in.

Gully control planting (using willows or poplars) is sometimes carried out at strong points such as a confluence upstream above an eroding gully head. In these situations, the areas should be fenced off to ensure that the plantings become well established. Once the plantings are established, careful grazing management can be carried out, taking care that the trees are not damaged.

13.3 EROSION CONTROL FORESTRY ON EARTHFLOWS AND OTHER DEEP-SEATED MOVEMENTS

Deep seated mass movements have been successfully stabilised by erosion control forestry on both large scale eg Mangatu Forest, and small scale, as with farm conservation woodlots. Research carried out at Mangatu Forest in Pinus radiata forests on argillite earthflows demonstrated the dramatic effect that the forest had on ground water levels. In addition there are significant soil reinforcing effects produced by the deep vertical sinker roots that develop about 7 years after planting.

At Mangatu the entire ground surface was covered, but where only discrete earthflows are being planted further action may be needed to ensure maximum effect on groundwater. Even at Mangatu there were areas where a combination of high groundwater (possibly due to subsurface flow) and poor soil permeability created trees with very shallow roots. These trees were subject to severe windthrow once the area was opened up by adjacent logging.

The chances of success may be improved by additional measures:

- De-watering of the earthflow before planting may increase stability in the early years after planting, before the trees are fully effective. It may also reduce windthrow problems in small blocks with a greater proportion of exposed edge trees.
- Ridge tops and areas of disturbed ground above the actual earthflow may be infiltration zones that feed groundwater into the earthflow. These may need minor earthworks to increase run-off and reduce infiltration, and should possibly also be planted to increase moisture loss. Soil water control will increase as greater proportions of a catchment are planted.
- In argillite country groundwater is known to flow between adjacent catchments, and this increases the importance of groundwater control where

only discrete catchments are to be planted. If obvious infiltration zones exist in neighbouring catchments they may need treatments as detailed above.

- Areas that are obviously wet and areas in which continuing root strength after felling is required, could be planted in water tolerant coppicing species such as redwood or suitable varieties of poplar or eucalyptus. The choice of tree may be influenced by erosion control forestry subsidy refulations.
- Roading should be carried out carefully to ensure that run-off is not concentrated or directed into critical areas, such as zones of permeable rock or gully heads, where it may increase gully instability.
- Planting densities for P radiata and Douglas Fir are recommended at 1500 stems per hectare in order to obtain root zone canopy closure for maximum early stability improvement (1250 sph under the ECFP 2001 regulations). Of equal significance is the subsequent silviculture regime as excessively zealous thinning too early can open the canopy and reduce the erosion control properties of the forest.
- The Gisborne District Council recommend thinning to 600 sph after the low prune and a final stocking of 350 to 400 sph. An alternative used by forestry companies producing un-pruned farming grade timber is to thin to 350 sph after 7 to 9 years, which is satisfactory form a soil erosion control perspective.
- Poplar may be planted at 500 stems per hectare under the ECFP regulations. An initial density of 1000 sph, using stakes rather than poles, would obtain more rapid root reinforcement.

13.4 EROSION CONTROL FORESTRY OF LARGE MUDSTONE AND ARGILLITE GULLIES (Farm Wood lots)

Standard erosion control forestry practice applies where entire catchments are being forested. Where only the gully itself, and some surrounding land, is being treated as a farm wood-lot, the limited understanding of the effects of forest on gully activity needs to be appreciated.

The reduction of groundwater flow to the gully walls, through afforestation of the gully's catchment, is almost certainly a major stabilising factor. Seepage forces and pore water pressure in a gully wall increase the incidence of the shallow slump failures that lead to gully enlargement. A reduction in the volume of groundwater that seeps from further up-slope will be beneficial and the greater the proportion of catchment planted the greater the benefit will be.

Evapotranspiration and interception by the forest will probably be of greater importance than the root strength of trees, except for those trees immediately adjacent to or on the gully walls. A reduction in surface run-off through the gully bed, is a further process by which afforestation aids gully stabilisation.

A less recognised role of the trees around the rim of a gully is their effect on the microclimate in the gully itself. Shading and wind protection of exposed faces by larger trees can greatly improve the possibility of them being revegetated.

13.4.1 Gully perimeter planting

The larger the proportion of catchment planted, the greater the reduction in run-off. A compromise on the location of the fence around the gully will occur if both grazing and stability are desired. On more unstable gullies a wide margin (at least 30 metres, but preferably much more) is required, while more stable gullies in mudstones may be fenced closer. Economies of scale will apply when harvesting takes places and a larger area may give a greater return in terms of both stability and dollars.

Fencing costs may be minimised if a reliable, well constructed, electric fence is used and if cattle-only grazing is carried out on the remainder of the paddock during critical periods of tree establishment.

Appropriate tree varieties can be best judged from other local experience or the NZ Farm Forestry Association may provide information. While P. radiata is almost universally successful for planting around gully perimeters, some varieties of Eucalyptus and Acacia are better suited to specific regions and conditions. Cypress species, in particular Cupressus macrocarpa and C. lusitanica may have limited uses but only where soils are well drained, as they cannot tolerate high water tables. Their foliage may cause abortion in cattle. Details on eucalyptus, cypresses, and other species, for specific sites are given in the Plant Materials handbook for Soil Conservation, Volume 2.

Tree planting densities will be as for normal production forestry practice (1500 stems/ha for P.radiata is typical) unless under an agro-forestry regime. One can expect a less rapid erosion control benefit from wider spaced agroforestry trees. The silviculture regime practised on farm scale gully wood-lots may be affected by the large proportion of edge trees, the degree of exposure to winds and the need to maintain a relatively dense tree stocking for erosion control benefits.

13.4.2 In-gully planting

Planting in an active gully will often only be successful when the surrounding forest is several years old and has already reduced groundwater flow. There seems to be little point in attempting to revegetate the gully floor when fresh sediment is still being deposited and signs of improved stability, such as the establishment of pampas grass seedling, may be useful as a guide as to when tree planting should commence. Rather than having two age classes of timber trees, the use of suckering trees that will survive logging damage may have merit on gully walls. Robinia pseudo-acacia has been observed growing well in such sites and Acacia melano-xylon has also been proved useful in tough conditions, although initial establishment can be difficult. A. dealbata is an exceptionally drought tolerant tree, particularly during its establishment phase, but is has still to be proven in the acid soils found in many argillite gullies. Other Acacia varieties were tested in Mangatu Forest by FRI, who will still have information available. In acid sulphate affected gullies in Northland the Coral Tree (Erythrina x sykesii) has been found to be very easy to establish, although it is best suited to frost free areas.

Pines appear to be hardier than poplars in the tough conditions found in gully walls, but if suitable suckering poplars can be established, such as Bolleana (Populus alba "Pyramidalis") or the P. alba x glandulosa crosses; Yeogi 1 and Yeogi 2 (which require well drained sites), they may provide long term stability despite logging damage.

The walls of argillite gullies will almost certainly be more difficult to revegetate than mudstone. In either case repeated attempts may be required until benevolent weather conditions occur and the plants are able to establish. The presence of seed sources of native plants such as Tutu, will often allow natural regeneration to occur once the major gully movements have been controlled by the surrounding forest.

A technique for establishing poplars and willows on steep slopes in some soil conditions is brush layering, in which cutting sized material is laid in a crossed pattern on contour terraces across the slope, which are then back filled with material from the next terrace up-slope. Most commonly used on road cuttings, the technique would only be warranted where very good soil moisture was present and the soil had adequate fertility. Details can be found in Schiechtl, 1980.

13.4.3 Post Harvesting Management

Where suckering trees and shrubs have become established in the gully floor and walls, provision should be made to exclude stock immediately after harvesting to allow these plants to recover from damage sustained during logging, prior to forest re-planting.

13.4.4 References

Schiechtl Hugo Meinhard. 1980. Bioengineering for Land Reclamation and Conservation. (English-language translation co-ordinated by N.K. Horstmann). University of Alberta Press.

Forestry company manuals, where these are able to be accessed, may provide standard methods of forest establishment.

After 1989 when Regional Councils were formed, and Government subsidies were no longer available for soil conservation activities, financial assistance with erosion control forestry became the responsibility of each Council. Some responded with ratepayer funded grants and others set aside capital funds from the sale of assets from which dividends were invested in conservation works. Others entered into joint ventures with farmers, where eventual proceeds from the sale of logs would be shared between the Council and the landowner.

13.4.5 Application

Erosion control forestry began on the east coast of the North Island with the plantings on Mangatu in 1961. The 6,500ha area was subject to slumps and earthflows, with severe gullying and aggradation in stream channels. The land was waterlogged in winter. The only vegetation consisted of poor pasture and a scattering on titoki and manuka (Willis 1973)

Early plantings were made with radiata at 2200 stems per hectare and by 5 years canopy closure had occurred and there was a marked improvement in slope stability. Observations showed that the top 10 to 20cm was much drier and there was a noticeable reduction in sub surface flows into stream channels. Aggradation had slowed and many of the streams and gullies had degraded to a stable base.

Current practice is to plant seedlings at a density of 1250 spha in 4m rows at 2m centres, with a thinning to about 400 spha while maintaining early canopy closure. (Marden 1995) With the ECFP it is possible to plant other species of trees, including exotic species such as Douglas fir and poplar, as well as indigenous forest plant. Much more is known about the effects of radiata on slump and earthflow erosion control due to the 40 years plus experience with this particular tree. With other species it is necessary to plant to similar tree densities in order to obtain similar erosion control results.

Seedling planting requires careful handling of the seedling tree from nursery to placing in the ground. Handling, transport to the site, preparing the ground, and eliminating plant competition is essential. Factors that should be considered are:

Handling and transport

After undercutting and wrenching in the nursery, the seedlings are lifted and placed in either plastic bags or cardboard boxes. There are advantages and disadvantages with each method of packaging. In boxes the seedlings are protected from crushing and sweating that might occur in plastic bags. However, with care, the bags can be more economical to transport. In the Gisborne area helicopters are sometimes used to transport men and seedlings into remote and difficult access sites. A plywood pod has been designed that can carry up to 3000 seedlings, and be slung under the helicopter. The pod is a 1.2 metre cube which is reusable and it provides good protection for seedlings in plastic bags.

Planting

The key to planting success is placing a well rooted plant in the ground where the roots are undistorted, facing downwards and contained by a well worked soil. Planting into a single spade cut with exposed roots and poor compaction raises the odds against tree survival. In drier areas a technique has been developed with considerable success where a double cut is made and the soil between the cuts is thoroughly loosened before placing the tree in the ground.

Reducing weed competition

It is normal forestry practice to use a herbicide to eliminate plant competition. A sprayed circle of 75 to 100cm is usual, the larger size is used where there is likely to be growth of vigorous pasture species. If the new tree is heavily overgrown by other plants it can lead to deformation and shortage of soil moisture, both factors essentially killing off the chance of a useful tree eventuating.

Other species

Erosion control forestry can include species other than radiata. Site conditions and accessibility may suggest the use of alternative species. For example, willows could be used where there is active gully erosion, macrocarpas for sheltered and accessible sites, and in along stream margins indigenous plants could be encouraged, and even planted.

Where poplars are selected the tree density needs to be much closer than normal open spaced planting.

Post planting care

Trees which are planted at 1200 to 1400 spha will need to be thinned to around 400 spha. The aim is to get early canopy closure, and thin out unwanted or distorted trees, while maintaining an even spread throughout the planted area. Pruning is a choice of the owner and a regime should be selected at or about the time of planting. Many factors will influence this decision, including, access to the site, proximity to processing, severity of erosion on the site and availability of time and resources to do the work

Erosion control forestry on Class VIIe land. *Photo: I H Cairns.*

13.5 FOREST MANAGEMENT PRACTICES – MOUNTAIN LANDS

13.5.1 Introduction

The practice involves planting a range of seedling trees on mountain faces or gullies subject to erosion. The trees are normally planted as seedlings but in special cases may be planted poles or stakes

Forestry plantings have been traditionally established to control slips, gullies, debris avalanches and to a lesser extent for frost heave, sheet wash, wind and creep erosion. There was very little planting in the mountain lands prior to the 1950s partly due to emphasis on other forms of revegetation and farm development but also there was insufficient information about the performance tree species at high altitude.

However, the work of Forest Research Institute at Craigieburn and Kaweka has provided valuable information on appropriate tree species, planting techniques, growth rates and other management details.

13.6 PROTECTION FORESTRY

This applies where forestry plantings are carried out with the objective of controlling soil erosion. The trees may in time have other benefits such as timber production, weed suppression, recreational uses, water quality enhancement and carbon storage.

13.6.1 Species

Many species have been tried in the 130 years of European settlement. The most successful are the conifers. Other species e.g. Poplars, Willows, Silver birch (Betula verrucosa) and sycamore (Acer pseudoplatanus) are less common in the high country.

- The most common species recommended today are as follows:
- Mountain pine: (Pinus mugo) very hardy, multiple leaders, shrubby. Most suitable for the severe high altitude sites up to say 1600m depending on the site.
- Lodgepole pine: (Pinus contorta) very hardy and seeds well but not recommended today because of wilding spread.
- Scots pine: (Pinus sylvestris) not used so much today, may be prone to possum damage.
- Corsican pine: (Pinus nigra) a hardy species (up to 800m) best on drier sites.
- Ponderosa pine: (Pinus ponderosa) often grouped with P.nigra as a species for hard sites but is more prone to pests and diseases when stressed.
- Radiata pine: (Pinus radiata) the most versatile quick growing species but is limited to the lower altitude or warmer sites below 540 m above sea level.

- Douglas fir: (Pseudotsuga menziesii) species well suited to the more moist areas of the high country.
- European larch: (Larix decidua) was used in the past but site tolerance and slow growth make it not as popular, deciduous, also potential wilding spread.
- Alders: (Alnus spp) still useful on moist sites esp on scree slopes and streambanks. Deciduous but can grow at altitudes up to say 1600m.

Other species are available such as the natives (normally slow and difficult to establish unless self seeded) and some hardy eucalypts (subject to frosting and pest problems).

13.6.2 Planting regimes

Generally for erosion control purposes seedling trees need to be planted at 1200-1600 stems/ha i.e. 2.5m x 2.5m to 3 x 3 m spacing. Follow up grass control is essential once the seedlings are planted but on many of the high country sites competitive grass and weeds may not be an issue.

On heavy rainfall areas where soil fertility status is low the addition of fertiliser tablets for each tree is recommended (trace element deficiencies occur in this environment). On the easier slopes that are readily accessible, follow up weed and pest control is essential. Silvicultural management needs to be considered, depending on the objective of erosion control.

Where frost heave is active severe difficulties may arise with seedling tree stock, so an alternative is to aerial oversow Pinus species seed with fertiliser and a small quantity of suitable grass-clover mix to reduce the frost lift. Otherwise, use Maku lotus, which is a species that adapts well to low fertility sites. Time sowing when moisture is present and stock have been removed from the block.

Where the trees listed above have been planted for erosion control purposes on areas with slight to moderate sheet, wind and creep erosion, they can be managed to enable harvest for income. Harvesting must be carried out using cable techniques or special logging equipment (e.g. Wyssen Hauler) for sensitive areas. The need for a detailed harvesting plan is paramount. Refer to chapter on slips for more information on harvest of trees from erosion-prone land.

Chapter - 14

Shelterbelts

Field shelter by planting windbreaks is widely practised on both agricultural and horticultural croplands and pastoral farmland. Many windbreaks were planted and fenced for grazing control and were often planted as part of regional wind erosion control schemes, which could attract a subsidy of up to 70% (Stringer 1978; Wethey 1984). By 1984, wind erosion control schemes protected 1.9 million hectares of land susceptible to wind erosion (Salter 1984). Windbreaks have been multipurpose, providing livestock shelter and enhanced crop growth (Sturrock 1981), in addition to mitigating wind erosion. Considerable research has been directed at windbreak design and performance, suitable species, the advantages of shelter, and its costs and benefits (Gilchrist 1984, Sturrock 1984).

In the "Plant Materials Handbook for Soil Conservation" 1986, Volume 1, p. 125-138, Allan Wilkinson gives a brief but comprehensive summary of windbreak siting, design, establishment and management, and examples of windbreak designs, both for shelter only, and as "timberbelts" for wood production. He emphasises the need for an integrated shelter system for a farm property. His format is followed here. Section 14 of the "Radiata Pine Growers' Manual" (McLaren 1994) contains additional information specific to pine timber belts.

14.1 SITING

The following are the major factors to be taken into account:

- Orientation should be as near as possible to 90° from the harmful wind direction.
- Windbreaks should be as long as possible, connect with existing shelter to obtain continuity, and gaps should be avoided where possible, because wind velocity is increased by up to 25% around ends and through gaps.
- The interval between windbreaks depends on the expected maximum height and degree of shelter required.
- Physical restrictions need to be taken into consideration, including property boundaries and roads, power and telephone lines, microwave and cellphone towers, water races, irrigation systems (especially border dyking), filed drainage systems such as tile drains and perforated pipes (novaflo), and existing fence lines.

- Existing natural shelter. On hill country, windbreaks should be complementary to natural shelter.
- Aesthetic considerations. Where possible, windbreaks should follow natural boundaries, and should include designs and species that enhance the existing local character; they should form part of an overall shelter system for the farm which includes hedges, woodlots and wide-spaced tree plantings.
- The farmer's objectives. These may be many, including minimising width of windbreaks, provision of shelter at a specified time of year (e.g. lambing, spring flowering, autumn crop and fruit harvest), minimising winter shading, minimising competition with adjacent crops or pasture, maximising the return from timber sales, increasing the pollen and nectar supply for honey bees, and providing food and shelter for desirable bird species [and insect predators].

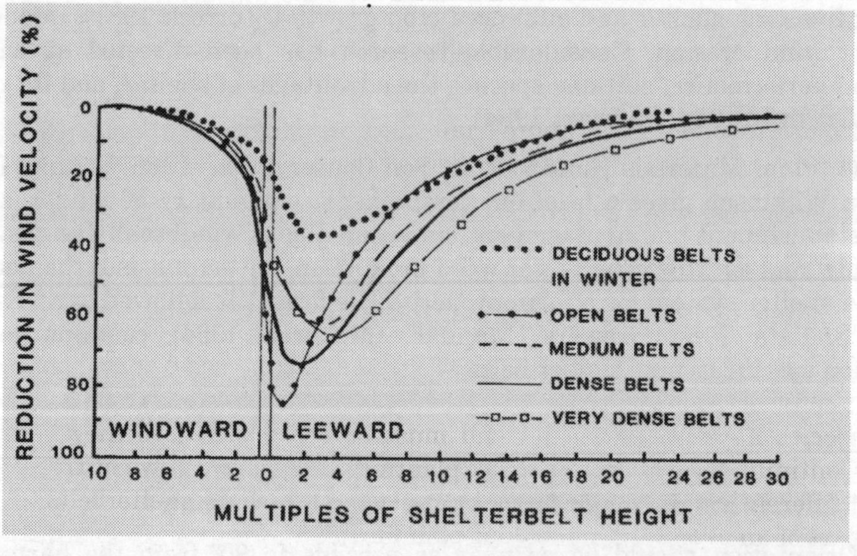

Patterns of wind abatement in the vicinity of shelterbelts of different density (after Caborn 1965).

14.2 DESIGN

Some weighting of objectives is necessary to produce designs that meet the needs of the farmer and the rural community. In the past, lack of perception and insufficient effort devoted to design has resulted in a limited number of species being planted throughout New Zealand without due regard to the local conditions and the visual landscape. Unreasoned antagonism to shelter has arisen in many areas. Single purpose designs such as radiata pine hedges have evolved and the contribution which shelter can make, both in increased production and the quality of rural life, has not been fully recognised.

The requirements for each windbreak site within a designed shelter system should be carefully examined. The resulting windbreak design should incorporate experience resulting from previous designs, the best local knowledge of species performance, and consideration of the following factors:

Height of windbreak

A significant reduction in wind velocity is obtained for a distance of 10-15 times the height of the windbreak. The maximum reduction occurs at a distance of 1-4 times the height, depending on the permeability.

Permeability (density)

Very dense impenetrable windbreaks can result in severe turbulence in the lee of the shelterbelt and a resumption of unhindered wind velocity at a relatively short distance form the shelterbelt. Permeable shelterbelts provide a greater reduction in wind velocity further out from the shelterbelt but less reduction very close to the shelterbelt. The optimum density for a permeable windbreak is between 40% and 50%. This may be obtained by:

- Planting trees of the appropriate natural density, e.g. pines have a dense crown, gums and poplars have light crowns;
- Varying the spacing of trees within the row;
- Varying the number of rows of trees [multiple rows];
- Tending, including side trimming, pruning, and thinning.

Continuity of shelter

All species have a limited life and must be replaced before they become over-mature. Continuity can be obtained by planting at least two rows of trees, each row having a different rotation length, or by replacing alternate shelterbelts at different time, as part of an overall management plan for the property.

Stability of windbreaks

Windthrow occurs in areas with shallow soils, which experience high velocity winds. The risk of windthrow can be decreased by:

- Good design, e.g. two rows of species which have different growth rates are combined to give additional stability to the windbreak: the slower growing species is established on the windward side to reduce movement of the root plate of the fast growing species;
- Good establishment techniques, e.g. the use of high quality tree stocks with some form of deep cultivation and careful planting;

- Good management: the milling and replacement of trees nearing the end of a rotation is the most significant step that can be taken to reduce future windthrow.

14.3 ESTABLISHMENT

For the establishment of shelterbelts, it is extremely desirable that 100% survival be obtained. Thus the greatest possible attention must be paid to site preparation, planting techniques, tree stock quality, fertiliser requirements, release spraying, protection form animal damage, and irrigation. The following are key points to note. (See Chapter 2 of the Plant Materials Handbook for more detail.)

14.3.1 Site preparation

Some form of cultivation is recommended prior to planting. The site can be cultivated well in advance (including deep ripping) and sown down in a covercrop such as oats or ryegrass to provide initial shelter for trees planted in sprayed spots or lines, and to suppress weeds. Shallow line ripping can be carried out in winter with at least two or three rips per line to avoid any possibility of root runs down single ripped lines. The tractor wheels can be run back over the rips to consolidate the soil and improve the surface for preplant herbicide application.

14.3.2 Time of planting

On sites where frost and waterlogging are not a problem, autumn planting is often very successful, especially if summer droughts are common. Wet or frosty sites are best planted in September-October if a period of stagnation in growth is to be avoided. Many North Island sites can be planted any time from May until late August. Many sites in the South Island and some in the central North Island cannot be planted until after the danger of severe frost damage is past.

14.3.3 Planting stock quality

Planting stock quality has a major effect on survival, early growth rate, root system configuration and resistance to windthrow. Container grown stock generally shows a higher survival and faster early growth rate than bare-rooted stock. However, where adequate survival and growth of bare-rooted stock can be obtained, this is preferred.

14.3.4 Weed control

Although good weed control can be obtained short-term by cultivation prior to plantings, more encouraging long-term control can be obtained by the use of herbicides. Only a limited range of herbicides is suitable for use in windbreak establishment because of the susceptibility of many of the species used (particularly hardwoods) and the need for 100% survival of good healthy trees. Accurate calibration of spraying equipment is essential to obtain successful weed control without damage to the trees. Residual herbicides commonly used include hexazinone, simazine, and terbumeton/terbuthylazine (see Appendix 2 of the Plant Materials Handbook, Vol 1).

Those farmers who do not wish to use herbicides, should consider various mulches, including black plastic mulches, or spraying with hot water or even silage leachate, or cultivation with the old push hoe.

14.3.5 Damage from rabbits, hares, and possums

The susceptibility of young trees to animal damage necessitates the use of individual protection devices, rabbit netting, or electric fencing (see Section 2.10.2 of the Plant Materials Handbook).Damage can also be reduced by:

- A programme of eradication of animal pests in conjunction with the regional council;
- Good establishment to enable trees to attain a size out of reach of animal pests as quickly as possible;
- Spot spraying well in advance of planting to allow the growth of tall grass around the planting spots before planting. (All three pests are less likely to run through rank grass than along clear-sprayed lines).
- Several Regional Councils also suggest pest repellent made by mixing together 5 fresh eggs, 150 ml of white acrylic paint, and 600 ml of water. Apply to seedlings with a paintbrush.

14.3.6 Windbreaks

Windbreaks should always be protected from grazing animals. The distance from fence to trees for areas carrying only sheep should not be less than 1 metre. Where cattle are present, the minimum recommended distance from the fence to the trees is 2.5 m. Where this distance to the windbreak is not acceptable, electrified top wires are recommended, but the farmer should ensure that this is a fail-safe electric fence. Animals have an uncanny knack of getting into a shelterbelt as soon as the power is off where they can do a lot of damage

14.3.7 Irrigation

Trickle irrigation is recommended on dry sites to ensure 100% survival and optimum growth rates. The amount of water to be applied depends to a large extent on the plant species, weather and soil type. Average application rates vary from 10-25 litres per tree every 7-10 days; light sandy soils require more frequent application than loam or clay soils. Additional advice may be sought from regional councils' irrigation officers or companies supplying the equipment.

14.3.8 Management & Maintenance

While some species for shelterbelts require little maintenance once well established, shelterbelts do require some maintenance to prevent them from:

- Spreading out beyond the boundary fence;
- Becoming too tall and casting excessive shade;
- Becoming too dense, causing excessive turbulence behind the shelterbelt.

Thinning may be applied, but grazing of the vegetation inside the fence is not recommended. This is to prevent the formation of a browse line under which windspeed could increase, or loss of beneficial insect predators

14.4 WINDBREAK DESIGN AND PLANT SPECIES FOR DIFFERENT SITES

Wilkinson l.c. describes in considerable detail, a range of shelter designs from one row to multiple rows hardwoods which can be deciduous or evergreens, with or without underplanting; also rows of conifers, again with a range of variations. The conifers may or may not be used for timber. He also describes shelter for coastal situations. The shelter designs are described in terms of shelter height, spacing, management, site selection and applications. The site tolerances of the suggested shelter species are indicated in Appendix 1 of the Plant Materials Handbook for Soil Conservation, Volume 1. Additional information may be found in the Fact Sheets of the New Zealand Plant Materials Research Collective. Advice about locally suitable species can be sought from Regional Councils, Landcare Groups and members of the Farm Forestry Association.

In all following figures, the wind direction is always from the left in the profile diagrams.

14.4.1 Shelterbelt, medium-tallhardwood, one row

Medium height species	Tall species
Alnus cordata	Populus alba 'Pyramidalis' (suckers)
glutinosa	
incana	P. X euramericana 'Tasman' or 'Veronese' or 'Crow's Nest'
rubra	
Phebalium squameum (syn. P. billardieri)	P. nigra 'Italica'*
	Salix matsudana**
Salix discolor	S. matsudana X alba hybrids**
	S. matsudana X pentandra hybrids**

* P. nigra 'Italica' is rust-susceptible; use only in inland Canterbury, Marlborough and Otago.

** Salix matsudana and its hybrids may be defoliated by larvae of the willow sayfly.

Initial Spacing

1-2 m between trees for horticultural shelter

2-3 m between trees for farm shelter

Management

Low maintenance. Optional side trimming to reduce width of belts and retain live green branches on the lower portion of the stem.

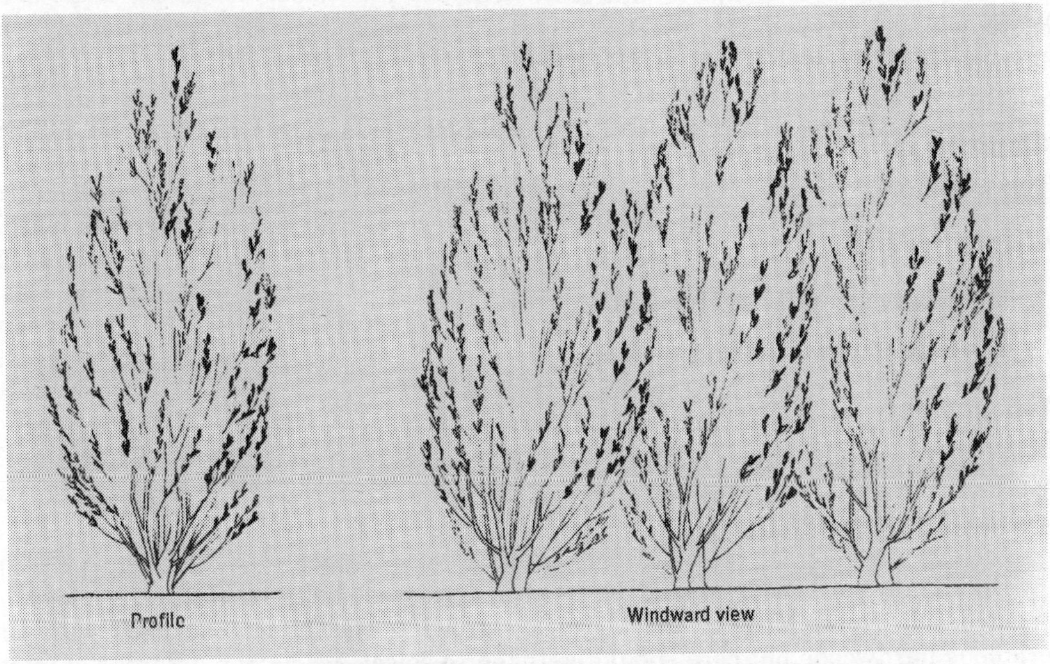

Medium-tall hardwood, one row.

Site selection/applications

Moist, reasonably fertile soils and irrigated cropland. Branch trimmings can provide supplementary stock fodder during mid-summer.

Use deciduous shelter for east-west windbreaks where winter shading is a problem.

Preference is now for perimeter belts to be of poplars and willows; internal belts of alder; in milder climates, the semi-evergreen Alnus acuminata.

14.4.2 Shelterbelt, medium-tallhardwood, one row, underplanted

Low- or slow-growing evergreen species	Tall, fast-growing species
Acacia spp.	Tall species listed in 14.4.4 or Eucalyptus spp. Listed in 14.4.4
Bambusa oldhanil	
Callistemon spp.	

Chamaecytisus palmensis
Corokia spp.
Cortaderia spp.
Cryptomeria japonica
X Cupressocyparis
Olearia spp.
Phormium cookianum
P. tenax
Pittosporum spp.
Thuja plicata

Initial planting

2-3 m between tall-growing species

1 m between low-growing species

Management

Low maintenance. Occasionally side trim to reduce fence overhang.

Site selection/applications

Deciduous hardwoods should be used in east-west belts on naturally moist or irrigated soil types. Most of the above low-growing species can be used with the deciduous hardwoods but care should be taken to select Acacia spp. tolerant of moist sites, e.g., Acacia floribunda, A. melanoxylon, A. retinodes.

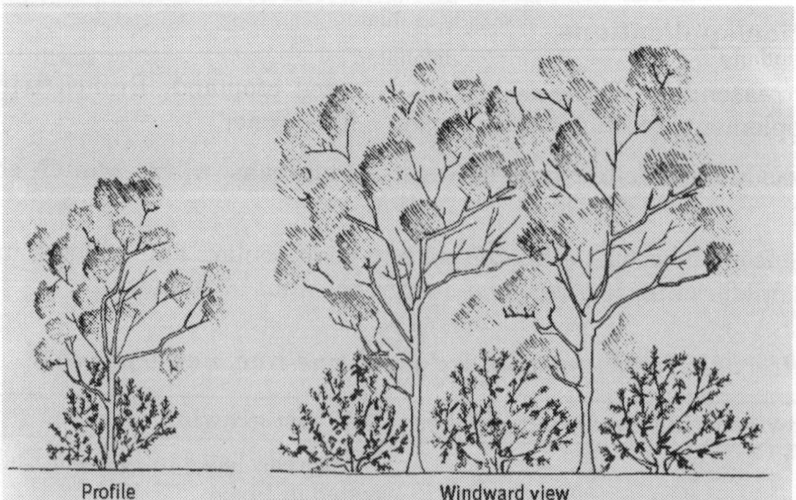

Medium-tall hardwood, one row, underplanted.

On drier sites use eucalypts combined with drought-tolerant, low-growing hardwoods or slow-growing conifers.

Hardwood shelterbelts can provide firewood, visual amenity, nectar and pollen for honey bees, and branch trimmings for supplementary fodder (deciduous species and Chamaecytisus). In milder climates fruit or nut-bearing shrubs such as feijoas, guavas and hazelnuts can be used for underplanting.

14.4.3 Shelterbelt, medium-tallhardwood, deciduous, two row

Low- or slow-growing species	Medium-tall growing species
Acacia floribunda	Alnus cordata
A. melanoxylon	gluthosa
A. retinodes	incana
Callistemon spp.	rubra
Chamaecytisus palmensis	Populus alba
Cortaderia spp.	'Pyramidalis' (suckers)
Cryptomeria japonica	P. X euramericana
Cupressus lusitanica	'Tasman'* or 'Veronese'*
X Cupressocyparis leylandil	or Crows Nest
Molaleuca spp.	P. nigra 'Italica'**
Olearia spp.	Salix matsudana***
Phormium cookianum	S. matsudana
P. tenax	X alba
Pittosporum spp.	S. matsudana
Thuja plicata	X pentandra hybrids***

*'Tasman' and 'Veronese' can be pruned for veneer logs. If this is done, prune annually to restrict stem diameter over stubs to 100 mm.

**P. nigra 'Italica' is rust-susceptible; use only in inland Canterbury, Marlborough and Otago

*** Salix matsudana and its hybrids may be defoliated by larvae of the willow sawfly

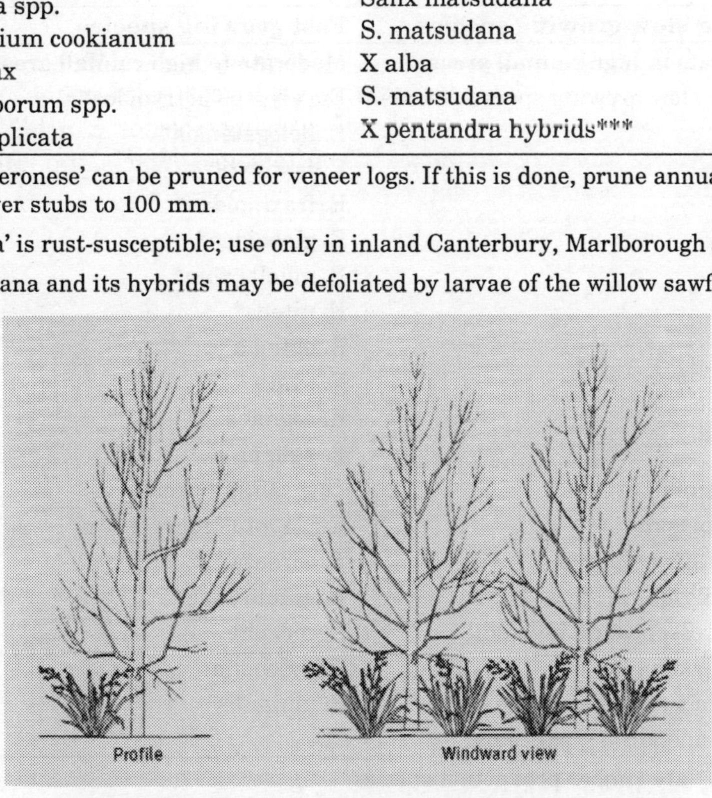

Medium-tall hardwood, deciduous, two row

Initial spacing

2-3 m between tall-growing species within the leeward row

1-1.5 m between low-growing species in the windward row

1-1.5 m between rows

Management

Wider-crowned species may require occasional side trimming.

Site selection/applications

Moist, reasonably fertile sites and irrigated cropland. Provides low winter shelter for stock; high permeable shelter for pasture and crops during the growing season. Use as east-west windbreaks where winter shading is a problem. Low-growing species and Salix spp. provide food sources for honey bees. Branches of Chamaecytisus palmensis and poplar and willow tree species can be used as supplementary fodder.

14.4.4 Shelterbelt, medium-tallhardwood, evergreen, two row

Low-or slow growing species	Fast growing species
Moderate in high rainfall areas.	Moderate to high rainfall areas.
Low or slow growing species listed in 14.4.3	Eucalyptus botryoides*
	E. delegatensis*
	E. fastigata*
	E. fraxinoides*
	E. globoidea*
	E. muellerana*
	E. nitens*
	E. obliqua*
	E. ovata
	E. regnans
	E. saligna
Low rainfall areas	Low rainfall areas
Abies pinsapo	Euclayptus amygdalina
Cupressus arizonica	E. cordata
C. torulosa	E. gunnii
X Cupressocyparis leylandil	E. nicholii
Pinus ponderosa	E. pulchella
P. nigra subsp. Laricio	E. vininalis
Sequoiadendron giganteum	

Species marked * are timber production species.

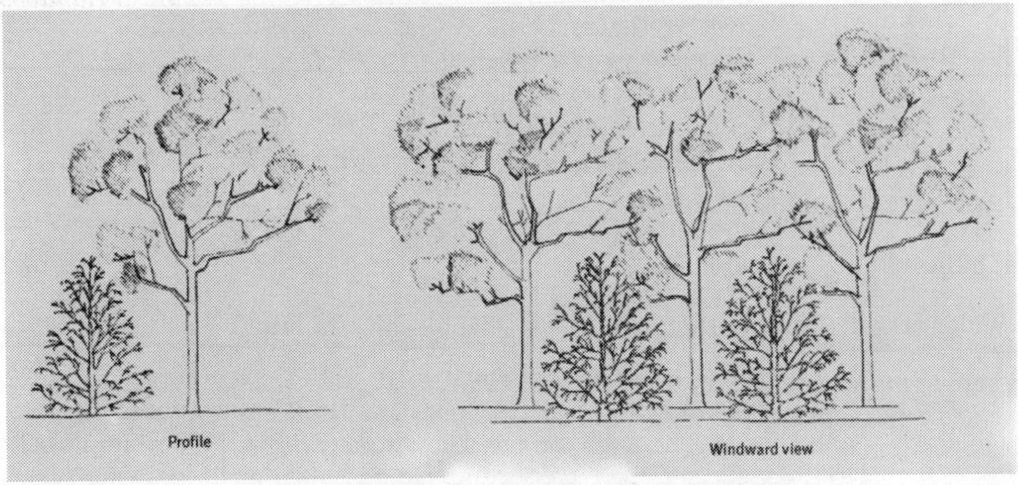

Medium-tall hardwood, evergreen, two row.

Initial spacing

4 m between trees in the leeward row

1.5-2 m between trees in the windward row

2 m between rows

If fast initial shelter is required, plant the leeward row at 2 m between trees; thin to 4 m apart at age 4 6.

Management

Low maintenance. Side trim occasionally to restrict width of belt. In shelterbelts the timber species require annual pruning to restrict stem diameter over stubs to 100 mm and to remove multiple leaders.

Applications

Timber, firewood, visual amenity, and nectar and pollen for honey bees and birds. Low-growing species can provide supplementary stock fodder from branch trimmings. It is essential to select the species and seed sources suited to particular sites.

14.4.5 Shelterbelt, conifer, one row

Fast-growing species
Pinus radiata
P. muricata (blue strain)
Medium-growing species
Cryptomeria japonica
Cupressus lusitanica

C. macrocarpa
X Cupressocyparz s. leylandii
Pinus nigra subsp. laricio
Pseudotsuga menziesii

Slow-growing species

Abies pinsapo
Cedrus atlantica
C. deodara
Cupressus arizonica
C. torulosa
Sequoiadendron giganteum
Thuja plicata

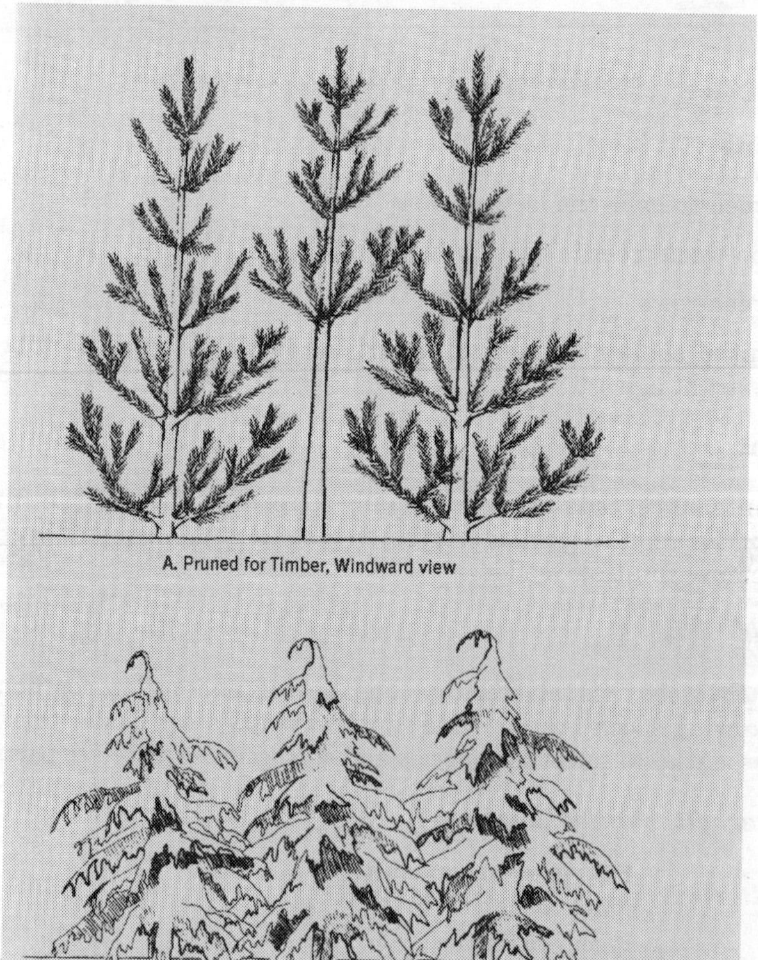

Conifer, one row.

Initial spacing

2-3 m between trees

Management High maintenance: Clear prune alternate trees (annually); pruning lifts should concentrate on restricting stem diameter over pruned branch stubs to 100 mm. Fan prune and/or side trim remaining trees. Fan pruning entails removing branches that grow at 30-90' to the windbreak. Prune logs to at least 6 m above ground.

Low maintenance: Optional side trimming to reduce the width of the windbreak.

Site selection/applications

One-row pruned belts are not suitable for primary shelter in very exposed coastal or inland South Island locations.

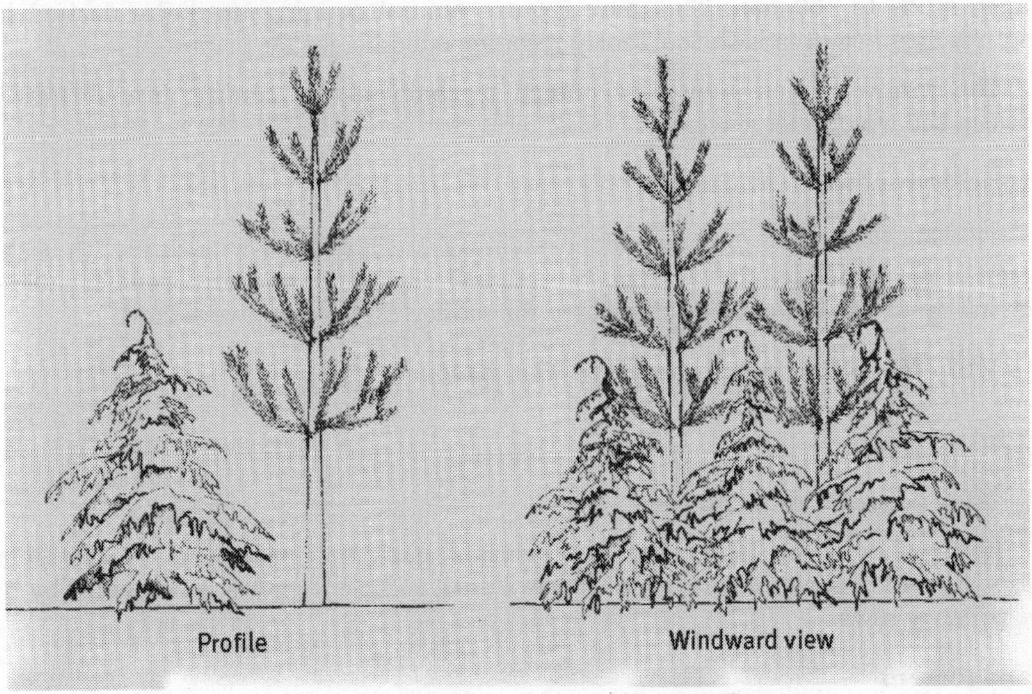

Tall conifer, two row, timber.

14.4.6 Shelterbelt, conifer, two row, timber

Slow-growing species	Fast-growing species
Low rainfall areas	Pinus nigra subsp. Laricio
Abies pinsapo	P. ponderosa
Cedrus deodara	P. radiata

Cupressus arizonica
C. torulosa
(2) Medium-high rainfall areas
Cryptomeria japonica
Pseudotsuga menziesii
Thuja plicata
X Cupressocyparis
Hardwood species listed in 14.4.3

Initial spacing

2-3 m between trees and rows

Management

Clear prune all trees in the leeward row to restrict stem diameter over pruned branch stubs to 100 mm. This may require annual pruning until the desired log length is attained (6 m is the currently recommended height for pruning).

The windward row may be trimmed mechanically to confine branch growth between the windbreak fences.

Site selection/applications

The windward row provides added stability (resistance to windthrow) thus this design is recommended for shallow soils. If desired, evergreen hardwoods capable of growing up to 6 m tall can be substituted for conifers in the windward row.

14.4.7 Shelterbelt, conifer, multiple row, timber

Initial spacing

2-3 m between trees and rows

Rows should be staggered to overlap gaps in preceding rows. Delay establishment of slow-growing leeward rows until sufficient shelter is provided by the two primary rows.

Management

Clear prune second (and subsequent rows if desired) to restrict stem diameter over stubs to 100 mm. Multiple row timber plantings should be pruned and thinned as woodlots. Amenity rows can be pruned to 3-4 m to allow underplanting with shade tolerant shrubs for fodder, pollen, nectar and fruit production.

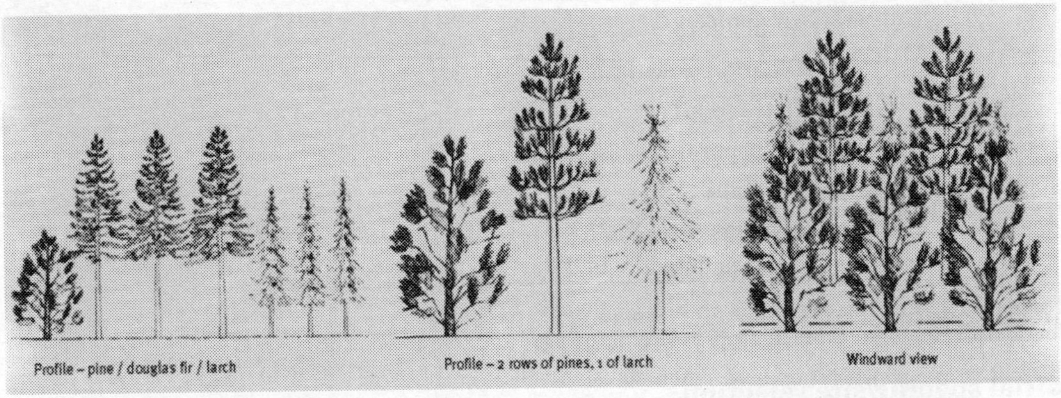

Tall conifer, multiple row, timber, cold inland areas of the South Island.

Application

Cold areas of the inland South Island where drifting snow is a major problem. Homestead shelter.

14.4.8 Shelterbelt, low-medium coastal, two row

Windward species

Acacia sophorae*

Coprosma repens

Cortaderia spp.

Phormium cookianum

P. tenax

Senecio reinoldii

Tamarix chinensis

Teucrium fruticans

Leeward species

Albizzia lophantha

Araucaria heterophylla

Banksia integrifolia

Cordyline australis

Corynocarpus laevigatus

Dodonaea viscosa

Erythrina sykesii

Lagunaria patersonii

Metrosideros excelsa

Myoporum laetum

M. insulare

Olearia paniculata

O. traversii

Pittosporum crassifolium

P. ralphii

Pomaderris apetala

Quercus ilex

*In many regions, this species is an invasive weed.
Check with Regional Council staff before planting.

Initial spacing/site selection

1-2 m between rows and between plants within rows

Spacings need not be rigid. Species can be grouped together in clumps to vary the width of the windbreak. Use the flax (Phormium) or toetoe/pampas (Cortaderia) with cabbage trees (Cordyline) on poorly drained heavy soils. In frost-fee areas Norfolk Island pine (Araucaria heterophylla) can be planted at 5-8 m intervals in the leeward row. Species for the South Island should be selected for greater cold tolerance.

Low-medium coastal shelter, two row.

Establishment and management

It may be advantageous to provide initial shelter for the plants by erecting a brush fence or commercial windbreak cloth barrier along the windward boundary of the shelterbelts. Maintain stockproof fences.

Applications

Low shelter for stock, visual amenity, shelter and food sources for birds and bees.

14.4.9 Shelterbelt, tall coastal, two row

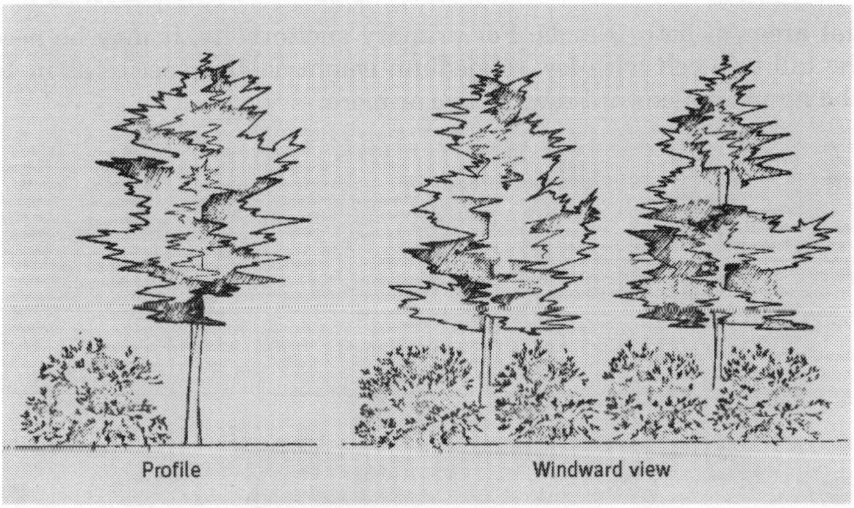

Tall coastal shelter, two row.

Windward species
Acacia sophorae*
Coprosma repens
Cortaderia spp.
Phormium tenax
Senecio reinoldii
Tamarix chinensis
Leeward species
Cupressus macrocarpa
X Cupressocyparis leylandii
Eucalyptus botryoides
Pinus muricata
P. radiata

*In many regions, this species is an invasive weed.
Check with Regional Council staff before planting.

Initial spacing

2.5-3 m between trees

2 m between rows

Management

Low maintenance. Occasionally side trim to restrict overhang on leeward side of windbreak.

Applications

Coastal areas of both islands. For primary shelterbelts, it may be necessary to precede the tall tree belt with low to medium height shelter species as in 14.4.8 and increase the number ofleeward rows to two or more.

Earthworks

Chapter

Chapter - 15

Runoff Control Practices for Earthworks

15.1 DIVERSION CHANNELS AND BUNDS

15.1.1 Description/Purpose

Diversion channels are shallow open drains which intercept and convey runoff to stable outlets at non-erosive velocities. They are used to protect work areas from upslope runoff and to divert sediment laden runoff to appropriate sediment retention systems. Bunds are compacted ridges of soil which intercept runoff and convey it to diversion channels.

15.1.2 Installation

Diversion channels and bunds need to be designed to ensure that they have sufficient capacity to convey the runoff from the required storm. They must have a positive grade to a stable outlet. Grade should be less than 2% otherwise the channel itself will erode and will then need to be stabilised (with rock, timber or fabric). The outfall may need to be protected against erosion rock, a flume. Because of the small room for error, these channels should be surveyed to obtain an even grade.

They should be constructed with a trapezoidal cross section with internal side slopes no steeper than 3 horizontal: 1 vertical. Earth bunds should be constructed in layers no greater than 200 mm in thickness and be well compacted as construction proceeds. As a general principle, any excavated channel or bund wall should be sown with grass seed or hydroseeded to prevent scouring.

Design requirements will vary from region to region and can be found in planning documents, erosion and sediment control guidelines etc. If there is no guidance, it is suggested that design be for the 1 in 20 year storm event.

Because of their low grade, deposition of sediment will often occur in diversion channels quickly compromising their capacity. Channels will need to be sited so deposited sediment can be removed. Care is therefore needed in their placement. If the channel/bund is constructed on a steeper grade to avoid deposition, the channel itself will generally erode (if constructed in earth) and will require specific measures to control this i.e. rock armour or fabric lining. A common problem is inadequate compaction of bund walls, or channels/bunds put in by eye, which breach or overtop at weak areas or low spots.

15.1.3 Maintenance

Diversion channels need to be inspected during and immediately after heavy rain, to ensure the diversion is working correctly and is not blocked or its function otherwise impaired. Any necessary repairs need to be undertaken. The outfall should be checked to ensure that it is free from erosion. Check that machinery movement through the site hasn't damaged the channel/bund.

15.2 CHECK DAMS

15.2.1 Description/Purpose

Check dams are small structures constructed across a drainage way or channel. Their purpose is to reduce erosion of the channel by reducing the velocity of flow. Although these measures also trap sediment, this is generally a secondary function.

15.2.2 Installation

The catchment of individual check dams should be limited to four hectares. They can be constructed from rock, logs, fabric, or sandbags and must have sufficient mass to withstand the energy of a concentrated flow. A check dam should be no more than 1000mm in height and have sides that are at least 300mm higher to ensure that any flow passes over the centre of the structure and not around the sides (which will usually wash it out). The toe and sides of the structure should be well protected against undercutting or outflanking. They should be constructed so that the crest of one structure is level with the toe of the next upstream (this can be hard to achieve in practice).

Fabric structures should be not more than 0.6 metres high, anchored well at the toe and well tied back. They are less robust and their catchment should be less than 2 hectares.

15.2.3 Limitations

Check dams are often undercut or outflanked and the failure of one can affect the integrity of other structures.

Diversion band.

Contour drains.

These are not commonly found on temporary earthwork sites because they take time to implement and can be difficult to construct correctly. They can be costly.

15.2.4 Maintenance

Check dams need to be inspected regularly after every storm. The inspection should check that the integrity of the structure is not impaired by erosion or piping. Sediment must be removed sufficiently often to ensure drainage channels behind each dam retain their flow capacity.

15.3 CONTOUR DRAINS

15.3.1 Description/Purpose

These are temporary drains constructed at the end of work or when rain is imminent to reduce slope length and convey runoff safely off an earthwork site. They are removed/obliterated once work recommences.

15.3.2 Installation

They should be as short in length as possible and constructed at no more than a 2% grade (to avoid scour of the channel). As a rule of thumb they should be about 500 mm deep and constructed at about 30 metre intervals. They should be closer together on steep sites whereas spacing can be increased on flatter sites. They should be as short as possible and constructed at no more than 2% longitudinal grade (to avoid scour of the channel). The position of these measures should be determined by the presence (or otherwise) of an erosion proof outfall. They must never discharge over areas of fill (install a flume).

15.3.3 Limitations

Care is needed with cut-off drains because they concentrate site runoff. Each drain must discharge to an outfall that is stable against erosion. Common problems include insufficient capacity, erosion prone outfalls, constructed on too steep a grade (so they erode) or too shallow a grade (so they don't flow).

15.3.4 Maintenance

They should be inspected during periods of prolonged rainfall to ensure they are functioning. Immediately after each heavy rainfall, any necessary repairs should be undertaken.

15.4 FLUMES

15.4.1 Description/Purpose

Flumes convey runoff down a bank or over an erodible site without causing erosion. Pipes, polythene, geotextiles, fabric flumes, sheets of iron, concrete lined chutes, rock etc can all be used for this purpose. Fabric flumes can be bought "off the shelf" (made to order from 200 – 600 mm diameter).

15.4.2 Installation

Flumes should be sized according to individual site requirements. The flume should be capable of carrying a minimum discharge equivalent to the 10% AEP (Annual Exceedence Probability) strom where the storm duration is equal to the time of concentration. Some flume dimensions for specific discharges are given in the reference "Gully erosion control techniques for pumice lands" Eyles (1993). On critical sites protecting valuable assets, the design strom can be as large as 2% AEP.

Plastic flume.

The catchment of temporary flumes should be a maximum of 2 hectares. All catchment flow needs to be directed to the flume. This is often achieved through a pipe installed through a compacted earth bund. Its invert should be laid at ground level and have a positive slope of at least 3%. All flow needs to be directed to the flume and the flume should be securely attached to the inlet pipe. The flume needs to be water tight, and the outfall should be protected against erosion. The flume should be laid directly down the slope if possible and be well supported and anchored. It should discharge to a stable outfall (rocks armour or erosion control fabric).

A secondary flow path may be required should the inlet become blocked or if flows exceed its capacity. Fish passage may need to be provided if a flume is to convey stream flows around temporary works in their channels.

15.4.3 Limitations

The inlet to the flume needs to be well dug in or protected (e.g. with concrete) to guard against undercutting or outflanking of flow. The wing wall, inlet and flume need to be well secured to each other. Wing walls may not direct all runoff to the flume. The flume may not be waterproof or runoff may splash out of the flume and erode adjacent areas. Subsidence may occur on fills and this may affect the functioning and integrity of the structure.

15.4.4 Maintenance

Flumes should be inspected daily during periods of prolonged rainfall, immediately at the finish of each rainfall, and weekly during periods of no rain. Any necessary repairs should be made immediately. The flume should be retained until the area has been permanently stabilised or the flow has been removed.

Chapter - 16

Soil Management Techniques on Earthworks

Stabilisation of an earthworks site commonly involves grassing, mulching, or fabric. Less common are all manner of structural measures, aggregate, various chemicals etc. Only the first three are discussed here.

16.1 GRASSING

16.1.1 Description/Purpose

Established grass protects thc soil against raindrop, sheet and rill erosion.

Application

Topsoil should be spread over the site to a minimum depth of 100mm to create a suitable medium for the establishment of grass seed. Grass seed should be applicd in either spring or autumn when there is enough moisture for germination and establishment but is not too cold to inhibit growth. Mulch can increase the success of grass establishment into the drier summer period. Annual grasses will generally give quicker establishment and more vigorous winter growth than will perennial grasses. Fertiliser should be added. A grass seed and fertiliser mix suggested in one region is as follows:

Table 16.1 Mix of grass seed and fertiliser for Auckland

	Mix	Rate (Kg/ha)
Seed	Temporary	300
	Annual rye grass (e.g. Tama) and clover	Perennial - 120
	Permanent	browntop - 45
	Perennial rye, brown-top and red/white clover	clover - 45
Fertiliser	D.A.P.	240
	N.P.K.S. (18:20:0:2)	

Ref ARC TP 90

Different seed mixes will be required for different regions. Discuss this with a local seed merchant. Hydroseed (seed, fertiliser and paper or wood pulp sprayed on as a slurry) can be used. Hydroseed can also be used in the cooler months (provided mulch is also applied), on difficult slopes, and as a temporary cover directly onto subsoil surfaces.

Limitations

Heavy rain soon after sowing can wash seed away (this particularly applies to hydroseed which is not incorporated into the soil). Grass can take a while to establish and additional measures, such as mulch, may also be required to protect the site against erosion while the seed establishes, particularly in the drier summer months. Poor grass establishment can often be attributed to insufficient fertiliser application.

Maintenance

Seed may need to be resown after heavy rain, or dry weather.

16.2 MULCHING

16.2.1 Description/Purpose

This is the application of a protective layer of straw, hay, wood fibre, bark or similar to the soil surface to protect it against raindrop impact and shallow sheet flow. It will give instant protection once applied and so can be used for erosion control when grass growth is slow (winter or summer) or where immediate protection is required. Hay has its own grass seed source but can carry weed seeds. Straw lasts longer. Mulch can "nurse" the establishment of grass seed. Bare subsoil can discharge over six times as much sediment as from mulched subsoil areas (ARC 2000) and mulch on topsoil is much more effective again.

Figure 2(38) mulching Photo

Installation

Mulch can be temporary or permanent, applied with grass seed and fertiliser or without. Mulch should be applied at 30 mm loose thickness (measured at time of application) and be anchored by spraying a tackifier or crimping into the soil with discs. It can be applied by hand on small areas but mechanical "blowers" will give a more even spread over larger areas. Runoff control measures need to be installed before mulching.

Limitations

Mulching will not control erosion resulting from concentrated flows so appropriate control measures should be applied first. Hay and straw will break down over time and another application may be required. Insufficient mulch may be applied. Some tackifiers can be environmentally harmful. Wind can blow mulch away, from exposed sites, unless it is anchored with wire netting.

Mulching.

Maintenance

Mulched areas need to be inspected after high winds or heavy rain and mulch reapplied as necessary.

16.3 GEOTEXTILES

16.3.1 Description/Purpose

Geotextiles are fabrics used to stabilise runoff conveyance systems. They can be used to line temporary channels, spillways, outlets to culverts, under riprap etc. An enormous variety of geotextiles is sold by manufacturers' agents. Which to choose, is dictated by cost as much as site.

Installation

The fabric must be dug in at the inlet. An anchor trench 500-mm wide by 500 mm deep needs to be excavated across the top of the channel and up the sides. The fabric should be laid in the trench leaving the top 1.5 metre extending out of the trench. The fabric should be pinned at 0.3 metre centres along the trench, and soil then backfilled and compacted over the fabric. The top 1.5 metre is then folded back over the compacted soil and pinned at 0.3 metre spacing. The fabric is rolled out down the channel and pinned on a 0.5 metre grid pattern down the sides and length of the channel. The fabric needs to be continued to an erosion proof outfall.

Limitations

The big drawback to most fabrics is that they allow some water through. Unless the channel is very smooth, some water can permeate and scour under the fabric. Polythene is waterproof but difficult to fasten (it rips if it is pinned). Some alternative fabrics address this problem but they are more expensive.

Pinning/fastening of fabrics is often poorly done, particularly at the inlet end. All fabrics must be securely dug in at the top of the work and well fastened. The function of the fabric liner depends entirely on the fastening. Where shear stresses are high more expensive fabrics and better fastening will be required.

Maintenance

The fabric should be inspected regularly e.g. weekly and during and after eachheavy rainfall to check it is securely fastened at the inlet and down its length.

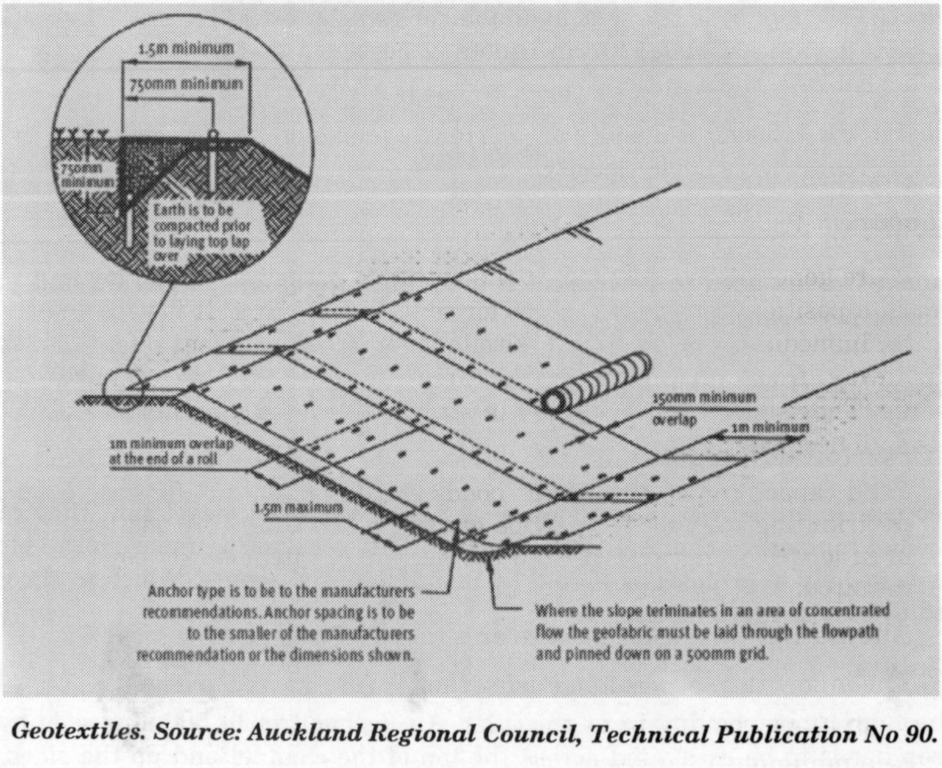

Geotextiles. Source: Auckland Regional Council, Technical Publication No 90.

Chapter - 17

Structures for Runoff and Sediment Control on Earthworks

17.1 SEDIMENT RETENTION PONDS

17.1.1 Description/Purpose

A sediment retention pond is designed to retain sediment on site from temporary construction and minimise off-site sedimentation. It is the most common sediment control measure utilised when the site is more than 0.3 hectares.

Installation

Sediment trapping efficiency is primarily a function of sediment particle size and flow velocity. To maximise the time available for sediment to settle out of suspension, a pond should retain as much storm runoff as possible, have low energy inlet flow, and low decant rate at the outlet.

The following design principles are currently promoted for clay soils. Silt or sand soils should require less pond capacity.

17.1.2 Location

These are generally constructed at the bottom of slopes, or at the lower end of a site, and in a position that allows access for sediment removal. It is usually better to locate them immediately below the works site (to optimise sediment retention), and, if possible, away from the actual works area in order not to compromise site earthworks. They should not be located in streams. Pond catchments should be kept to less than 5 hectares if possible.

- Pond capacity. The capacity of ponds should be 200 m3 for each hectare of contributing catchment for slope gradients less than 10% and less than 200 m in length. Otherwise 300 m3 per hectare of catchment of pond capacity is required.
- Alternatively: use Pond surface area (m^2) = 1.5 peak inflow rate (l/s).
- The inflow rate is calculated using the 5% AEP rainfall event.
- Pond depth. This should be between 1 – 2 metres in depth. The pond should be a minimum 1.0m deep.

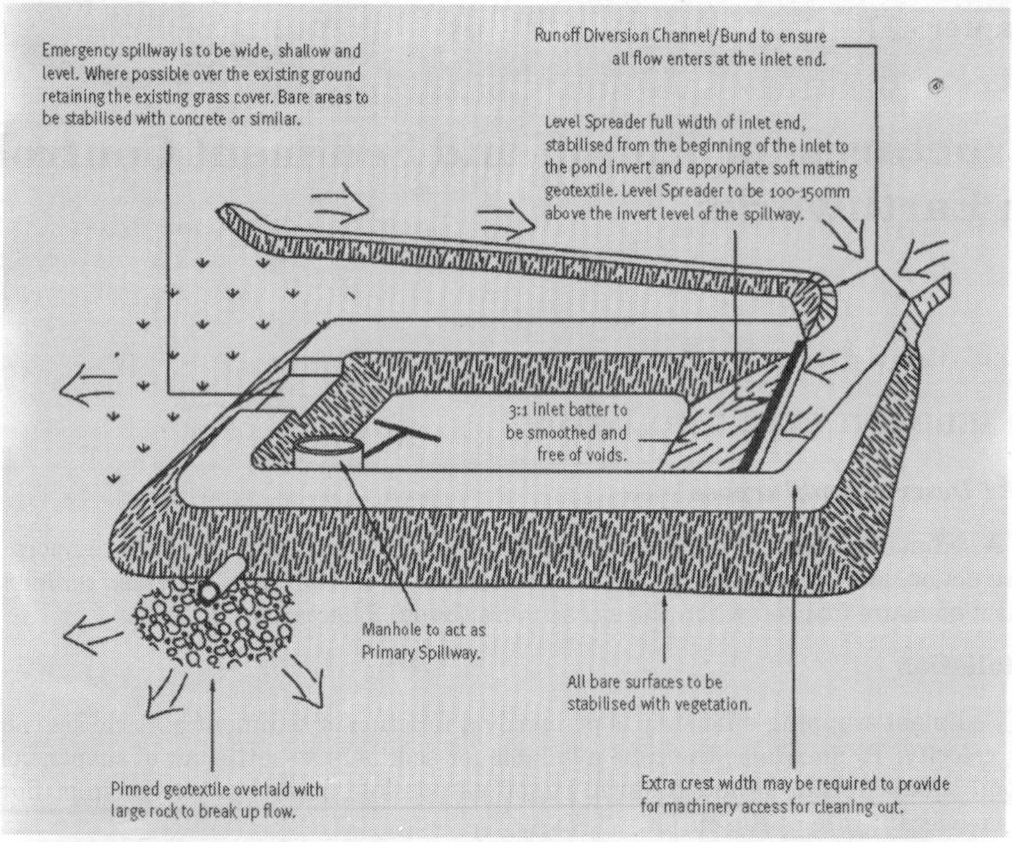

Sediment retention pond. ***Source: Auckland Regional Council, Technical Publication No 90.***

Sediment retention pond.

- Length to width. A pond should be at least three times longer than it is wide (baffles may be necessary to achieve this).
- Inlet. Inlets should be level, have an inlet batter of no more than 3 horizontal to 1 vertical slope, and be the same width as the pond.
- Primary Spillway. A piped outlet is required to act as the primary spillway. For ponds less than 3 hectares, the outlet pipe can be 150 mm diameter. Concrete manholes and pipes are required for larger catchments. At least two antiseep collars should be installed along the pipe through the pond embankment and the soil needs to be well compacted to guard against possible failure along the pipe.
- Emergency spillway. This should cater for the 1 % AEP rainfall event and be constructed in solid ground (not fill). Lining to protect it against erosion will be necessary. It should be at least 6 metres wide or the width of the pond (whichever is the greater).
- Permanent water storage. This is the component that remains in the pond to dissipate inflow energy and should be 30 % of the pond capacity.
- Operating volume. This is the volume between the lowest decant level and the primary spillway invert and should be 70 % of the pond capacity.
- Outlet decants. These should allow the removal of the relatively clean water only. The current recommended rate is 3 litres/sec/ha of contributing catchment and is achieved by a perforated floating decant pipe. This pipe is generally 100 mm in diameter and has a series of 10-mm holes drilled at 60 mm spacing along its length. This is attached to the primary spillway. One decant is required for each 1.5 hectare of catchment. A single decant should extend through the operating volume of a pond. If more than one decant is required, then the additional decants should be positioned so they "kick in" proportionally through the operating volume of the pond.
- Pond Cleaning. Ponds should be cleaned out when they are 20 % full of sediment.
- Pond efficiency. Ponds constructed in accordance with the above criteria are about 70-80% efficient in retaining sediment (from trials on clay based soils in the Auckland region). Increasing the pond size does not significantly increase the efficiency of sediment retention.
- Chemical treatment. Some trials with poly aluminium phosphate (PAC), alum and polyacrylamide have been undertaken to promote flocculation of suspended sediment particles and aid sediment retention. While results look promising, further investigation is being undertaken.

Limitations

The embankment and spillway can be weak areas. Careful attention should be paid to good compaction and erosion protection here. Settlement of fill may affect embankment heights. Pipes through embankments need antiseep collars and good compaction to protect against piping. Ponds are hazards and should be fenced.

Maintenance

The most common problem is decant holes blocked with floating material such as straw mulch. These should be checked daily. Ponds should be inspected on a regular basis and after every storm. Checks should be made for outflanking, scour, spillway protection structural soundness of the embankment and the operation of the decants. Repairs should be undertaken to ensure the pond remains in good working order. Sediment should be removed from the pond when it is 20 % full. The pond should be retained until the upslope area has stabilised.

17.2 SEDIMENT RETENTION BUNDS

17.2.1 Description/Purpose

A sediment retention bund is a compacted ridge of soil used to retain sediment from small areas of sheetflow only.

Installation

The catchment of a sediment retention bund should be less than 0.3 hectares. The bund should be well compacted for strength. Its outlet should have a perforated pipe upstand, about 100 mm below the spillway level, supported by a rigid stake/ waratah and connected to a non-perforated pipe that passes through the earth bund. The spillway should be lined with an impermeable or nonwoven fabric so excess flows can be safely discharged. The spillway should be the lowest point on the bund and a freeboard of 250 mm should be allowed for. The spillway should be stabilised e.g. lined with impermeable or non-woven fabric and this should be well pinned.

Limitations

They should never be used in waterways or where runoff has concentrated. Lack of compaction is often a problem, particularly if the discharge pipe is trenched through the embankment. Spillways on earth bunds are often poorly constructed.

Maintenance

They should be inspect regularly (e.g. weekly, during and after each heavy rainfall). They should be repaired or reinstated as necessary. Sediment should be cleaned out when the bund is 20 % full. The perforated upstand should be unblocked when it becomes clogged.

17.3 SILT FENCES

Description/Purpose

A silt fence is a geotextile barrier attached to posts and used to retain sediment from small areas of sheet flow only.

Installation

They are constructed from woven geotextile fabric. The supporting post/waratah should be at two metre maximum spacing and 400 mm into ground. The fabric should be trenched 200 mm deep into the ground and upslope of the supports. Gravel filled sandbags laid end to end over the toe of the fence can be used where trenching is not practical e.g. where there is surface rock. The top of the fabric should be 400 mm above the ground and have a tensioned wire (2.5 mm HT) along the top of the silt fence. The fence should be aligned along the contour as much as possible. Ensure that the fence cannot be outflanked by flow by turning the ends up equivalent to the effective height of the fence. Returns should be constructed every 30 metres to confine sediment load. The fence should be tied back to another waratah for additional support in low points.

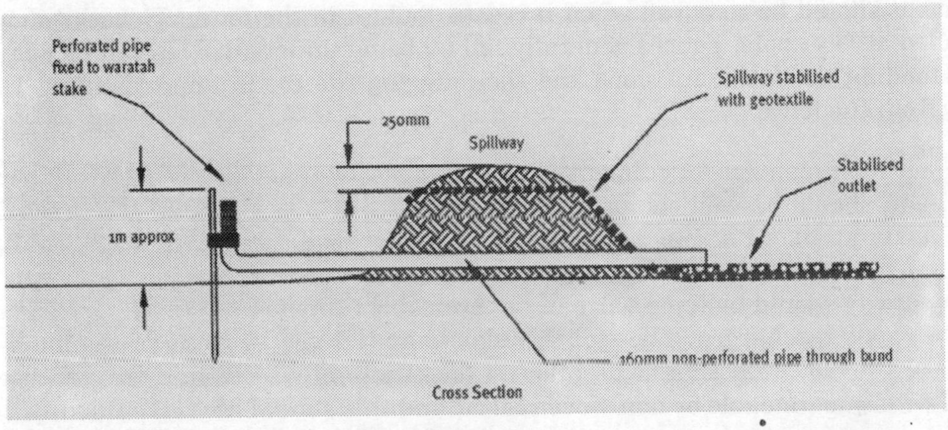

Sediment retention bund. ***Source: Auckland Regional Council, Technical Publication No 90.***

Silt fence

A silt fence can be reinforced with sheep netting, chain mesh netting or similar behind it. The netting must also be dug into the ground.

Limitations

Silt fences should only be used to intercept sheet flow. They should not be used across flowing watercourses or similar areas of concentrated flow, as they do not have the strength to stand the energy of concentrated flows. The fabric quickly becomes blocked with fines and the fence then acts more like a sediment retention measure. They often fail through being undercut.

Maintenance

They should be inspected after heavy rainfall and repaired as necessary. Sediment should be removed when it creates bulges in the fence. Tiebacks should be installed at low spots. Fences generally fail by being undercut. This can be addressed by extending the fabric up slope and then digging the toe in approximately 1 metre away from the fence

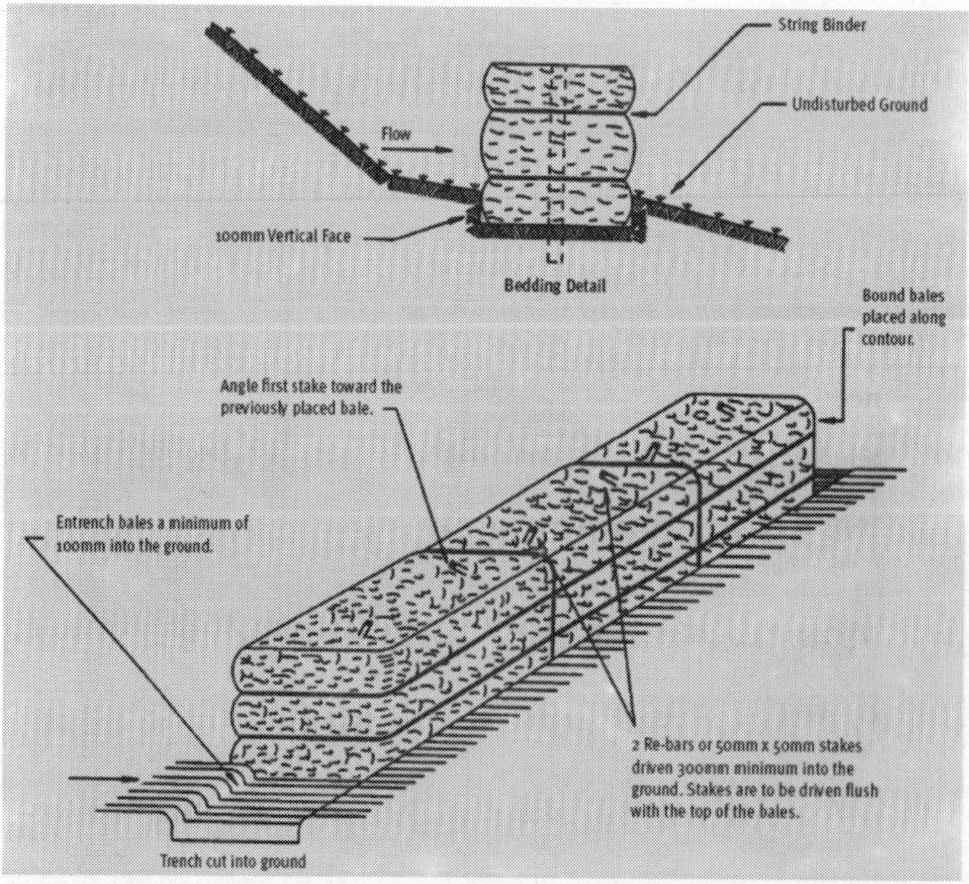

Haybale barriers. *Source: Auckland Regional Council, Technical Publication No 90.*

17.4 HAYBALE BARRIERS

Description/Purpose

A haybale barrier is a temporary sediment retention measure. They can also be used to intercept and divert runoff from small catchments.

Installation

The catchments should be less than 0.2 hectare for sediment retention purposes and 0.5 hectares for runoff diversions. The haybales should be placed on the contour and the ends of the barrier turned up to stop runoff flowing around them. Each bale should be butted up to the previous bale and each bale staked with 2 stakes driven 300 mm into the ground. The first stake in each bale should be driven towards the previously laid bale to force the bales together. Holes under/between the bales will need to be securely plugged with clay.

For runoff diversion purposes, start from the discharge point and then install the hay bales on a maximum 2 % grade. The line needs to be surveyed (an abney level or inclinometer is sufficient – do not survey by eye), and the hay bales will normally weave around a slope as the terrain varies. Make sure that the discharge point is secure against erosion, by installing erosion control fabric or rock armour.

Limitations

Hay bales are impermeable and do not filter. They should never be used in permanently flowing waterways or where runoff has concentrated. They have a short life of only a few months. Once they get wet, they become very heavy and difficult to move around on site.

Maintenance

For "tight"(sites, they can be removed after rain provided they are replaced before the next rainfall. Otherwise, leave them in place and inspect them regularly (weekly, during and after heavy rainfall). Sediment should be removed when it is half way up the bale. The barrier should be reinstated or repaired after each heavy rain event (or if it is damaged by site machinery).

Figure 2(45) Hay bales Photo

17.5 STORMWATER INLET PROTECTION

Description/Purpose

These are sediment retention measures installed around roadside stormwater inlets to capture eroded material, preventing its entry to stormwater systems and eventual discharge to receiving environments.

They can be constructed from sandbags, silt fences, rock/gravel barriers, or haybales. An excavated depression upslope of the measure helps retention.

Installation

The catchment of each protective measure should be less than 0.3 hectares. Individual measure should be spaced up slope of the stormwater inlet in the roadside water table. The impoundment depth should be slightly lower than the kerb to avoid redirecting the flow. Treated runoff should then flow around the outside of the sandbags and into the stormwater inlet. The stormwater inlet should not be covered with, silt fence fabric (because this will quickly block) but left open so flows can enter into the stormwater system. Consider whether ponded water will encroach on a road/private dwelling.

Rock should be 100 mm in diameter and faced with 25 mm gravel on the side away from the inlet. Side slopes should be a maximum of 3 horizontal: 1 vertical. Concrete blocks and wire mesh can be used as a support for the stone.

- Sandbags can be used as barriers – they should be half full only.
- Silt fences should be no more than 500 mm in height.
- Coarse filter fabric can be laid against wire mesh and both nailed onto a wooden frame secured by sandbags.

Limitations

Because flows have to be retarded to retain sediment, control measures can impede the entry of stormwater into stormwater inlet systems, and cause surface flooding. They should be regarded as a backup system only.

Maintenance

Control measures need to be repaired if dislodged or broken. They should be inspected, and sediment removed after each heavy rain.

Chapter - 18

Dust Control Measures for Earthworks

18.1 INTRODUCTION

Dust is generated when soil is repeatedly disturbed and broken down into fine particles. On susceptible soils (such as volcanic ash when dry, or loess) dust associated with earthworks can be severe, and difficult to control. Repeated tracking of soils with machinery not only breaks down the soil particles but also aerates them so that they become quite "fluffy" and suspended as particulate material in the air. This is similar in principle to sediment control where finer particles are more difficult to settle out. On high-risk sites, fine soil on the ground can become very dry and aerated, and roll in waves as machines pass even when there is no wind. Once the site is subject to any wind, the dust gets very difficult to control. Dust from problem sites can travel for kilometres and cause a range of problems to health and property.

18.2 GUIDELINES

Dust management should be considered early in the planning stages of any earthworks project. Planning and managing to minimise dust problems is the best option. If dust management is only addressed after it has become a problem, it is almost impossible to bring under control.

A Dust Management Plan should be prepared prior to any works being undertaken on a susceptible site. In the Bay of Plenty, where resource consents are required for some earthworks operations, a Dust Management Plan is a necessary part of the consent application.

The main practice used to control dust on earthworks is the application of water to keep soil moisture high enough to prevent dust generation. A Dust Management Plan should include the following elements:

- The potential effects of dust if it causes a nuisance off site.
- The soil characteristics of the site and whether the timing of operations will help or hinder dust control.
- Any methods that can reduce the dust e.g. restricting the amount of bare ground exposed, staging of works.

- If water is used, the plan should detail the water source, source capacity and availability. If the source is marginal, then on-site storage may be necessary. If the water is sourced from a municipal reticulated water supply, then written confirmation from the territorial authority will be required.
- Other types of control that may be used.
- Contingency plans (e.g. for severe wind problems). Contingency plans should outline other options if the primary method of control turns out to be ineffectual.

A Dust Management Plan normally also provides for signage at an earthworks site giving a 24 hour contact number for dealing with dust complaints that may arise from the operations. The provision of this 24 hour contact number ensures that the contractor has a management plan in operation to deal with dust control.

The timing of works can be crucial for dust management. If the earthworks can be carried out during wetter seasons, then dust control will be less of a problem.

18.3 WATERING

Description/Purpose

The application of water to maintain soil moisture so that the soil does not dry out sufficiently to generate dust.

Application

Water is normally applied for dust suppression in one of two ways; by water cart or by sprinkler. Either system requires a minimum amount of water to achieve effective dust control over an open earthworks site. In the Bay of Plenty, the minimum amount of water required to control potential dust problems is 5 mm/day.

The use of water carts is the most common system of dust control. Water carts can carry from 3,000 to 10,000 litres. The use of water carts is limited by the ability of the vehicle to access the areas that require wetting down. A sprinkler system is often used on earthworks sites where there are large areas open, or where the terrain may be too steep for water carts. Sprinkler systems are also commonly used on sites where some irrigation may be useful to establish vegetation following completion of earthworks.

Water should be applied as a dust suppressant at a rate of 5mm/day before soil moisture levels start to drop. This can vary with individual sites, and prevailing weather conditions, but consideration should be given to applying water after 7 days without rain during spring/summer periods on susceptible sites, This period can be reduced markedly in windy conditions, where on-site conditions should be assessed to commence treatment. Water should continue to be applied until the next rainfall event that results in surface runoff from the site.

Limitations

The main limitation is availability of sufficient water during mid-summer. When the water cannot be taken from municipal water supplies, alternative options may be required. Sometimes, a reservoir can be used.

The use of a groundwater bore specifically for the operation is a common practice.

18.4 DUST SUPPRESSANTS

Description/Purpose

Dust suppressants are polymers or chemicals applied to the exposed surface to protect it from the wind, and so reduce the dust generating capacity of the treated area.

Application

Dust suppressants are a recent alternative that has been used on some sites in New Zealand over the past two to three years. There is a range of dust suppressants available. They are normally a proprietary blend of naturally derived surfactants and acrylic polymers, provided in a aqucous emulsion form so that they dissolve readily in water for easy application. Upon application, they provide a protective surface that reduces dust emissions.

Alternatively, some dust suppressants act as a binder. These include salts (CaCl2, and MgCl2), and ligin sulfonates. Chemical suppressants include salts, lignin sulfonate, wetting agents, latexes, plastics, vegetable oils and petroleum derivatives. The use of petroleum derivatives may require a resource consent.

Limitations

Dust suppressants generally cannot withstand machinery being driven over the treated area. The cost means that they are only used in places where the treated area is not be worked for a period of time. There are questions regarding whether they have a contamination effect on the soil. The potential contamination effects are unknown at this time.

18.5 SURFACE STABILISATION

Description/Purpose

Protection of exposed ground surface using a range of different materials or products to reduce the potential for dust generation.

Application

Surface stabilisation is a dust control option that is normally only used as a last resort if other dust control measures are not fully effective. Hydroseeding mix, aggregate or geotextiles may be placed on to the exposed soil to reduce dust. Sometimes, aggregate just need to be applied on areas where there are a lot of vehicles travelling. Alternatively, aggregate may be applied in conjunction with a water cart as a back up.

Limitations

Surface stabilisation is normally only used as a last resort because of the its cost, and also because the surface stabilisation may then need to be removed to finish the job.

18.6 WINDBREAK FENCING

Description/Purpose

Windbreak fencing comprises low fences constructed from fabric to reduce wind velocities on site, and reduce dust generation.

Application

Wind break fencing using geotextile or windbreak fabric is sometimes used on small sites to help control particular areas that may be difficult to access using other methods, or to assist with the establishment of vegetation as an additional dust control measure. Windbreaks or silt fences can also be useful in keeping machinery off critical parts of the earthworks site.

Limitations

Windbreak fencing only controls dust on small areas because the fences are relatively low; no more than 1 metre high, and protect ground for no more than 10 metres to leeward. Even here, they can be ineffective due to eddies round either end of a fence. Unless well-staked and constructed out of heavy duty fabric, they will be demolished by strong winds.